Harald Neidhardt

Arsenic in groundwater of West Bengal: Implications from a field study

Impact of groundwater abstraction and of the organic matter on release and distribution of arsenic in aquifers of the Bengal Delta Plain, India

Karlsruher Mineralogische und Geochemische Hefte

Schriftenreihe des Instituts für Mineralogie und Geochemie,
Karlsruher Institut für Technologie (KIT)

Band 39

Arsenic in groundwater of West Bengal: Implications from a field study

Impact of groundwater abstraction and of the organic matter on release and distribution of arsenic in aquifers of the Bengal Delta Plain, India

by
Harald Neidhardt

Dissertation, Karlsruher Institut für Technologie (KIT)
Fakultät für Bauingenieur-, Geo- und Umweltwissenschaften, 2012
Referenten: PD Dr. Stefan Norra, Univ.-Prof. Dr. Thomas R. Rüde

Impressum

Karlsruher Institut für Technologie (KIT)
KIT Scientific Publishing
Straße am Forum 2
D-76131 Karlsruhe
www.ksp.kit.edu

KIT – Universität des Landes Baden-Württemberg und
nationales Forschungszentrum in der Helmholtz-Gemeinschaft

KIT Scientific Publishing 2012
Print on Demand

ISSN 1618-2677
ISBN 978-3-86644-941-1

Impact of groundwater abstraction and of the organic matter on release and distribution of arsenic in aquifers of the Bengal Delta Plain, India

Zur Erlangung des akademischen Grades eines

DOKTORS DER NATURWISSENSCHAFTEN

von der Fakultät für

Bauingenieur-, Geo- und Umweltwissenschaften
des Karlsruher Instituts für Technologie (KIT) – Universitätsbereich

genehmigte

DISSERTATION

von

Diplom-Geoökologe Harald Neidhardt

aus Karlsruhe

Tag der mündlichen Prüfung: 04.07.2012
Hauptreferent: PD Dr. Stefan Norra
Korreferent: Univ.-Prof. Dr. Thomas R. Rüde

Karlsruhe 2012

"Der denkende Mensch irrt besonders, wenn er sich
nach Ursache und Wirkung erkundigt: sie beide
zusammen machen das unteilbare Phänomen.
Wer das zu erkennen weiß, ist auf dem
rechten Weg zum Tun und zur Tat!"
Johann Wolfgang von Goethe

ABSTRACT

Naturally occurring arsenic-bearing groundwater threatens the health of millions of residents in the Bengal Delta Plain (BDP), which is amongst other Asian regions one of the most severe affected areas worldwide. For more than three decades, inhabitants have been exposed to arsenic-enriched groundwater, which is the often available source of presumably safe drinking water. This fatal misconception resulted in the widespread occurrence of chronic arsenic (As) intoxications. After more than two decades of intensive research, the interactions of biogeochemical processes induced by metal-reducing microbes and hydrological conditions have been identified as cause of locally high concentrations of dissolved As in shallow groundwaters. The present thesis aims at assessing the relative roles of biological and inorganic controls that underlie As mobilisation and accumulation at two study sites with contrasting As concentrations.

Based on a comprehensive sampling campaign in the Nadia district (West Bengal, India), an area well known for the occurrence of high As concentrations in shallow groundwater, two representative study sites were chosen. At each site, five nested monitoring wells were installed. The well screens cover different depth ranges and reach down to a depth of 45 m (low As site) and 37 m (high As site). During well drilling, sediment samples were taken in regular intervals of 0.65 m and the lithology was recorded. The lithology at both sites is similar. Fine and medium sandy aquifer sediments are capped by thin clayey and silty layers that form a surface aquitard. Here, mottles of secondary Mn- and Fe-(oxyhydr)oxides are accompanied by highest sedimentary As contents of up to 122 mg kg^{-1}. This depth range represents rather a sink for As than a source. Despite this, average sedimentary As contents of the underlying aquifer sands are characteristically low (high As site: 3.8 ± 1.2 mg kg^{-1}, low As site: 3.2 ± 1.3 mg kg^{-1}). Sequential extractions and statistical interpretations of geochemical element analysis revealed that Fe-(oxyhydr)oxides are the primary hosts for surface adsorbed as well as incorporated As.

Isotopic signatures of C and N were combined with C/N ratios determined from sedimentary organic matter to draw conclusions regarding the highly dynamic sedimentation history of the study area.

High concentrations of HCO_3^- (>400 mg L^{-1}), Mn (>0.35 mg L^{-1}), Fe (>1.10 mg L^{-1}), PO_4^{3-} (>1.10 mg L^{-1}) and the absence of NO_3^- and partially even SO_4^{2-} reflect pronounced microbial influences on the chemical composition of groundwater at the study sites, which resulted in moderate to strongly reducing conditions. Additionally, multiple and superimposed processes of groundwater evolution formed hydrochemically stratified water columns, particularly at the high As site. Characteristic As concentrations ranged from 98.0 to 296 µg L^{-1} at the high As site and varied between 49.1 and 155 µg L^{-1} at the low As site. The reduced form As(III) was constantly the prevailing As species.

After a one year long monitoring period, in-situ field experiments were performed. At the low As site, indigenous microbes were locally stimulated by infusing dissolved sucrose (saccharose) into the aquifer. This provoked a stimulation of Fe(III)-reducing microbes as indicated by Fe(II) concentrations in groundwater that temporary increased 36-folds. Although As concentrations concomitantly rose by up to 48.6 %, As mobilisation was decoupled from the reductive dissolution of metal oxides and the release of other trace elements with partly steep raises (Zn x 78.0, Co x 47.3, Ni x 36.5, and V x 33.1). The release of As into groundwater is primarily attributed to reductive dissolution of the hosting Fe-(oxyhydr)oxides, although an influence of As(V)-reducing microbes cannot be excluded. The decoupling of As release from microbial activity in general, and from Fe(III) reduction in particular, points at a fast adsorption of dissolved As to residual and/or newly formed Fe-minerals.

At the high As site, investigations focused on abiotic aspects of the As distribution in local groundwater. Here, extensive abstraction of shallow groundwater was simulated. The extraction promoted a rapid increase of As in the upper pumping well from 83.4 to 296 µg L^{-1}, which was caused by mixing with arsenic-enriched groundwater from deeper layers.

Between December 2008 and August 2010, all ten monitoring wells were sampled in regular intervals of two weeks. At the low As site, active As and Fe mobilisation occurred in the shallowest monitoring well (well screen located in 12-21 m below land surface), which was additionally super-imposed by seasonal fluctuations. These temporal trends were directly linked to vertical oscillations (~1.55 m) of the water table between the dry season and the monsoon rains. At the high As site, temporal trends in the local groundwater hydrochemistry appeared in one well, too. These changes were induced by clear pumping of a newly installed well (2008/2009) and by the groundwater abstraction experiment (2009/2010). Monitoring results further revealed that the hydrochemistry at both sites subsequently returned to the initial baseline values in the following weeks after the in-situ experiments were finished. The patterns of As release during the monitoring and the injection experiment strongly suggest that As release at the low As site was induced by Fe(III)-reducing microbes. Results further demonstrate that temporal fluctuations in dissolved As occur in shallow groundwater of the investigation area. Such temporary trends can either arise from active As release by microbial metabolic reactions and/or from vertical and horizontal flow of the hydrochemically stratified ground-water body. This flow originates from seasonal fluctuations of the ground-water table, but it can also be connected to groundwater abstraction.

The abstraction experiment clearly shows that pumping endangers the limited and still arsenic-free groundwater resources of the BDP. Results were discussed in context with findings of previous studies in an effort to develop an advanced concept of As release and distribution within the investigation area. The outcomes of this concept can be further transferred to the entire BDP as well as other affected Asian regions. This concept is based on the close relationship, which exists between As, Fe and PO_4^{3-} during water-sediment interactions and arises from microbial Fe(III) reduction. This induces in turn concomitant precipitation and transformation of Fe-minerals and competitive adsorption of dissolved As and PO_4^{3-}.

The active zone of As release and enrichment occurs in reducing aquifer parts in approximately 20 to 40 m depth, where a biogeochemically controlled environment of competing As release and retention prevails.

ZUSAMMENFASSUNG

Natürlich auftretende hohe Gehalte an Arsen (As) im Grundwasser gefährden die Gesundheit von Millionen Einwohnern der Bengalischen Delta Ebene (BDE). Diese ist, neben anderen Regionen Asiens, eines der weltweit am stärksten von diesem Phänomen betroffenen Gebiete. Seit mehr als drei Jahrzehnten sind die Bewohner arsenhaltigem Wasser ausgesetzt, welches die oftmals einzig verfügbare Quelle für vermeintlich sicheres Trinkwasser darstellt. Diese tragische Fehleinschätzung führte zu einer andauernden Exposition, was letztendlich zum massenhaften Auftreten von Symptomen chronischer Arsenvergiftung führte. Nach mehr als zwei Jahrzehnten intensiver Forschung konnten Wechselwirkungen zwischen biogeochemischen Prozessen (hervorgerufen durch metallreduzierende Bakterien), sowie die örtlichen hydrologischen Rahmenbedingungen als Ursache der hohen Arsengehalte in oberflächennahem Grundwasser identifiziert werden. Ziel der vorliegenden Studie ist es, die relativen Anteile biologischer und abiotischer Kontrollmechanismen an der Freisetzung und Anreicherung des Arsens am Beispiel zweier Standorte mit gegensätzlichen Arsengehalten zu ermitteln.

Basierend auf einer umfangreichen Analyse des lokalen Grundwassers wurden zwei repräsentative Untersuchungsstandorte im Nadia Distrikt des indischen Bundesstaates West Bengalen ausgewählt, welcher bekannt ist für das Vorkommen erhöhter Arsengehalte im Grundwasser. An beiden Standorten wurden jeweils fünf in unterschiedlichen Tiefenbereichen verfilterte Grundwassermessstellen eingerichtet. Die Filterstrecken decken verschiedene Teilbereiche des Aquifers ab. Im Falle des niedrig arsenbelasteten Standorts reichen diese bis zu einer Tiefe von 45 m, während der hoch arsenbelastete Standort bis in 37 m Tiefe verfiltert ist. Während der Installation der Messstellen wurden fortlaufende Kernproben von je 0.65 m Länge gewonnen. Dabei wurde zusätzlich die auftretende Lithologie erfasst, welche an beiden Standorten sehr ähnlich ist.

Die aus Fein- und Mittelsanden aufgebauten Grundwasserleiter werden jeweils von dünnen, tonigen und schluffigen Schichten überdeckt, welche einen an der Oberfläche anstehenden Grundwasserhemmer bilden. Dieser Bereich erscheint durch das Auftreten von sekundären Mangan- und Eisen-(oxyhydr)oxiden als marmoriert, wobei hier die höchsten sedimentären Arsengehalte von bis zu 122 mg kg^{-1} nachgewiesen werden konnten. Diese jüngsten Ablagerungen scheinen daher eher eine Senke denn als eine Quelle für Arsen zu sein. Im Gegensatz dazu sind die durchschnittlichen Arsengehalte der darunterliegenden Grundwassersedimente deutlich niedriger (hoch arsenbelasteter Standort: 3.8 ± 1.2 mg kg^{-1}, niedrig arsenbelasteter Standort: 3.2 ± 1.3 mg kg^{-1}). Sequentielle Extraktionen und statistische Auswertungen der Elementanalysen zeigten außerdem, dass Eisen-(oxyhydr)oxide die primäre Quelle für oberflächenadsorbiertes, sowie in die Kristallstruktur eingebundenes Arsen darstellen. Die Isotopen-signaturen (δ^{13}C und δ^{15}N) sedimentärer organischer Substanz wurden zusammen mit den C/N Verhältnissen dazu verwendet, die dynamische Sedimentationsgeschichte des Untersuchungsraumes zu rekonstruieren.

Durchgehend hohe Gehalte von gelöstem HCO_3^- (>400 mg L^{-1}), Mn (>0.35 mg L^{-1}), Fe (>1.10 mg L^{-1}), PO_4^{3-} (>1.10 mg L^{-1}) und die Abwesenheit von NO_3^- sowie stellenweise SO_4^{2-} zeugen von einem starken Einfluss mikrobieller Prozesse auf die chemische Zusammensetzung des Grund-wassers im Bereich der Untersuchungsstandorte. Dies führte zu moderaten bis stark reduzierenden Redox-Milieus im Aquifer. Mehrere, sich teils über-lagernde Prozesse der Grundwasserentwicklung erzeugten im Laufe der Zeit hydrochemisch stark differenzierte Wasserkörper, was besonders beim hoch arsenbelasteten Standort deutlich wird. Typische Arsengehalte variierten hier zwischen 98.0 und 296 µg L^{-1}, während die Gehalte des gering belasteten Standortes zwischen 49.1 und 155 µg L^{-1} lagen. Dabei stellte reduziertes As(III) die generell dominierende Arsenspezies im Grundwasser dar.

Nach einem Jahr regelmäßiger Probenahme wurden in-situ Feld-experimente durchgeführt. Beim gering mit Arsen belasteten Standort wurden indigene Bakterien innerhalb des Aquifers durch die Zugabe gelöster Saccharose angeregt. Dies bewirkte eine erfolgreiche Stimulierung

Fe(III)-reduzierender Bakterien, was durch kurzfristige, aber teilweise starke Anstiege der Gehalte an gelöstem Fe um das bis zu 36-fache des Ausgangsgehaltes verdeutlicht wurde. Obwohl die Arsengehalte im Grundwasser parallel dazu um bis zu 48.6 % anstiegen, erscheint die Mobilisierung generell entkoppelt von der reduktiven Auflösung von Metall-(oxyhydr)oxiden und den dabei freigesetzten Mengen anderer Spurenelemente, welche teilweise extrem deutliche Anstiege verzeichneten (Zn x 78.0, Co x 47.3, Ni x 36.5 und V x 33.1). Der Anstieg der Arsengehalte ist primär der reduktiven Auflösung von arsenhaltigen Fe-(oxyhydr)oxiden zuzuschreiben, wobei ein möglicher Einfluss As(V)-reduzierender Bakterien nicht ausgeschlossen werden konnte. Die Entkopplung der Arsenmobilisierung von der Intensität mikrobieller Aktivität im Allgemeinen, sowie der Freisetzung von gelöstem Fe im Speziellen, deutet auf eine sofortige Adsorption von gelöstem As an residuale, beziehungsweise neugebildete Eisenminerale hin.

Die Untersuchungen am stark mit Arsen belasteten Standort waren primär auf abiotische Aspekte der Verteilung von As im Grundwasser fokussiert. Hier wurde die massive Entnahme von oberflächennahem Grundwasser simuliert. Diese Entnahme verursachte einen raschen Anstieg der As Konzentrationen von 83.4 auf 296 µg L^{-1} im Grundwasser, welches der zentralen und gleichzeitig auch flachsten Grundwassermessstelle entnommen wurde. Änderungen in der hydrochemischen Zusammensetzung des Grundwassers sowie des Arsengehaltes konnten in diesem Bereich des Aquifers auf Mischungsprozesse geschichteter Wasserkörper zurückgeführt werden.

Zwischen Dezember 2008 und August 2010 wurden in den 10 Grundwassermessstellen in regulären Abständen von zwei Wochen Wasserproben entnommen. Bei dem gering arsenbelasteten Standort konnte im flachsten Beobachtungsbrunnen (verfiltert im Bereich 12 bis 21 m unter Geländeoberkante) ein Anstieg der Arsengehalte infolge aktiver Fe(III) Reduktion beobachtet werden, welcher durch saisonale Konzentrationsschwankungen überlagert wurde. Die saisonalen Trends gingen direkt mit Änderungen der Standrohrspiegelhöhen bis zu 1.55 m einher, die zwischen der Trockenzeit und des Monsunregens auftraten. Im Falle des stark mit

Arsen belasteten Standortes konnten ebenfalls zeitliche Schwankungen der hydrochemischen Grundwasserzusammensetzung in einem der überwachten Brunnen verzeichnet werden. Diese Veränderungen konnten auf die Simulation der Grundwasserentnahme (2009/2010), sowie das Klarspülen des zentralen Brunnens nach dessen Neuinstallation (2008/2009) zurückgeführt werden. Die Resultate der Langzeitbeobachtung zeigten außerdem, dass die direkten Auswirkungen der beiden Feldexperimente auf die lokale Hydrochemie auf nur wenige Wochen beschränkt waren.

Die beobachtete Freisetzung von As am gering arsenbelasteten Standort wurde sowohl während der Langzeitbeobachtung, als auch während des Biostimulationsexperimentes sehr wahrscheinlich durch die Stoffwechselaktivität Fe(III)-reduzierender Bakterien hervorgerufen. Die Ergebnisse dieser Studie zeigen außerdem, dass zeitliche Schwankungen der Arsengehalte in lokalem Grundwasser des Untersuchungsgebietes auftreten. Diese Schwankungen können einerseits durch die Tätigkeit anaerober Bakterien entstehen, andererseits aber auch durch horizontale wie vertikale Strömung im hydrochemisch geschichteten Grundwasserkörper. Eine solche Verlagerung kann wiederum auf saisonalen Schwankungen der Grundwasseroberfläche basieren, oder Folge einer intensiven Wasserentnahme sein.

Diese Ergebnisse zeigen deutlich, dass Grundwasserentnahme die begrenzten und bisher noch arsenfreien Grundwasserressourcen innerhalb der BDE gefährden. Die Zusammenführung der Ergebnisse dieser Studie mit bereits publizierten Erkenntnissen hat zu einem neuen Konzept der Freisetzung und Verteilung von As innerhalb des Untersuchungsgebietes geführt, welches auch auf die gesamte BDE sowie weitere betroffene Gebiete Asiens übertragbar ist. Die zugrunde liegende Hypothese dieses Konzepts basiert auf einer engen Verbindung zwischen As, Fe und PO_4^{3-} bei den auftretenden Grundwasser-Sediment-Wechselwirkungen und thematisiert die Freisetzung dieser Substanzen infolge mikrobieller Fe(III) Reduktion. Die Mobilisierung infolge von Fe(III) Reduktion ruft wiederum eine parallele Ausfällung und Transformation von Eisenmineralen hervor, was von einer konkurrierenden Resorption gelösten As und PO_4^{3-} begleitet wird.

Der aktive Bereich der Arsenmobilisierung und Anreicherung wird in reduzierenden Aquiferbereichen in ungefähr 20 bis 40 m Tiefe vermutet, wo eine biogeochemisch kontrollierte Umgebung mit konkurrierender Arsenfreisetzung und Zurückhaltung vorherrscht.

ACKNOWLEDGEMENTS / DANKSAGUNGEN

Die vorliegende Arbeit entstand am Institut für Mineralogie und Geochemie des Karlsruher Instituts für Technologie im Rahmen eines deutsch-indischen Kooperationsprojektes.

Mein besonderer Dank gilt Dr. Zsolt Berner für dessen unermüdlichen Einsatz für das Projekt. Insbesondere während der strapaziösen Arbeiten im Gelände war seine Erfahrung von unschätzbarem Wert. Die Tür zu seinem Büro stand stets offen und ich danke ihm für die zahlreichen Diskussionen die wir führten. Zsolt, ich hoffe du wirst deinen wohlverdienten Ruhestand entspannt und in vollen Zügen genießen!

Ebenso bin ich PD Dr. Stefan Norra zu großem Dank verpflichtet. Er hat ohne Zögern die Betreuung meiner Doktorarbeit übernommen, stand immer für Fragen und Diskussionen bereit und hat die zahlreichen Entwürfe meiner Artikel mit wertvollen Kommentaren versehen. Außerdem danke ich ihm für das immerwährende Vertrauen und die Unterstützung, die er mir schon seit meiner Diplomarbeit gewährt hat. Ich hoffe wir werden in naher Zukunft wieder zusammenarbeiten können.

Weiterhin möchte ich Prof. Thomas Neumann für die Unterstützung und die Möglichkeit danken, meine Dissertation in einem so positiven Arbeitsumfeld wie dem IMG erstellen zu können.

Ich bedanke mich bei Univ.-Prof. Thomas R. Rüde für die Übernahme des Ko-Referendariats.

ACKNOWLEDGEMENTS / DANKSAGUNGEN

Ich möchte mich an dieser Stelle bei der Deutschen Forschungsgemeinschaft und dem Bundesministerium für wirtschaftliche Zusammenarbeit und Entwicklung für die Finanzierung des deutsch-indischen Kooperationsprojektes bedanken, in dessen Rahmen diese Doktorarbeit entstand. Ich möchte hierbei auch den Antragstellern, Prof. Doris Stüben, Prof. Josef Winter und Prof. Debashis Chatterjee danken.

Selbstverständlich bin ich auch den Mitarbeitern des IMG und des IfGG zu Dank verpflichtet, die mir bei den umfangreichen Analysen, unterschiedlichsten technischen Belangen sowie Laborangelegenheiten aller Art jederzeit hilfsbereit zur Seite standen: Frau Claudia Mössner (HR-ICP-MS), Frau Cornelia Haug (Vollaufschlüsse), Frau Beate Oetzel (XRD), Herr Kristian Nikoloski (Dünnschliffpräparation), Frau Gabriele Konrad (technische Unterstützung), Frau Pirimze Khelashvili (IC), Dr. Gerhard Ott (Soft- und Hardwareunterstützung), Frau Gesine Preuss (IRMS) und Herr Martin Kull (C/N Analysen).

Weiterer Dank geht an / I further thank:

...   our Indian project partners from the Department of Chemistry, University of Kalyani: Prof. Debashis Chatterjee, Santanu Majumdar, Pabida and especially my friend Ashis Biswas for supporting the field work in West Bengal.

...   the people of Sahispur and Chakudanga, especially Mr Atul Mandal, for the custody we received during our field trips and the permission to install the monitoring wells.

...   den Projektpartnern des IBA, Prof. Josef Winter, PD Dr. Claudia Gallert und meinem Freund Dominik Freikowski für ihre Unterstützung des Projektes und die zahlreichen spannenden Diskussionen.

... Dr. Utz Kramar für die Unterstützung bei den Messungen an der ANKA und an der Micro-ED-RFA.

... Jörg Göttlicher und Ralph Steiniger für die Möglichkeit Messungen an der SUL-X Beamline der ANKA durchführen zu können.

... Dr. Elisabeth Eiche für das Korrekturlesen und die zahlreichen wertvollen Ratschläge, welche sie mir im Laufe der letzten Jahre gab.

... Nina, Alexander, Xiaohui, Maren, Daniel, Andreas, Farooq, Moritz, Steffi, Dirk, Peter, Alexandra, Jan und Monika für die großartige Zeit am Institut und die vielen unvergesslichen Momente.

... Dr. Martina Adamek für die Durchsicht meiner Entwürfe, das Ertragen meiner Launen und die unentwegte Unterstützung während den stressigen und schwierigen Phasen meiner Promotion.

... meinen beiden Eltern Theresia und Josef, denen ich diese Doktorarbeit widme. Ihre immerwährende Unterstützung ermöglichte es mir, ein derart anspruchsvolles Unterfangen wie eine Promotion zu einem erfolgreichen Ende zu bringen.

TABLE OF CONTENT

LIST OF ABBREVIATIONS AND SYMBOLS

ANKA:	Angströmquelle Karlsruhe
AOD:	Acid oxalate method under darkness
arrA:	Arsenate respiratory reductase gene
asl:	Above sea-level
AVS	Acid volatile sulphides
BDP:	Bengal Delta Plain
BGS:	British Geological Survey
bls:	Below land surface
BMZ:	Bundesministerium für wirtschaftliche Zusammenarbeit und Entwicklung (German Federal Ministry for Economic Cooperation and Development)
bp:	Before present
CAO:	Chemolitho-autotrophic bacteria
cfu:	Colony forming units
C_{org}:	Organic carbon
CSA:	Carbon sulphur analyser
DARPs:	Dissimilatory arsenate respiring prokaryotes
DCB:	Dithionite – citrate – bicarbonate method
DFG:	Deutsche Forschungsgemeinschaft (German Research Foundation)
DMAA:	Dimethylarsinic acid
DOC:	Dissolved organic carbon
DPHE:	Department of Public Health Engineering
EA:	Element analyser
EAWAG:	Eidgenössische Anstalt für Wasserversorgung, Abwasserreinigung und Gewässerschutz (Swiss Federal Institute for Environmental Sciences and Technology)
EC:	Electrical conductivity
EDX:	Energy dispersive X-ray spectroscopy
E_H:	Redox potential related to standard H_2 electrode
FeRB:	Iron reducing bacteria
GC:	Gas chromatograph
GI:	Gas ion
GISP:	Greenland Ice Sheet Precipitation
GMWL:	Global meteoric waterline
HAO:	Heterotrophic arsenide oxidising bacteria
HDPE:	High-density polyethylene
HR:	High resolution
IBA:	Institut für Biologie und Biotechnologie des Abwassers (Institute of Biology for Engineers and Biotechnology of Wastewater Treatment)
IC:	Ion chromatography
ICDP:	International Continental Scientific Drilling Program

LIST OF ABBREVIATIONS AND SYMBOLS

IfGG: Institut für Geographie und Geoökologie (Institute of Geography and Geoecology)

IMG: Institut für Mineralogie und Geochemie (Institute of Mineralogy and Geochemistry)

IPCS: International Program on Chemical Safety

IRMS: Isotope ratio mass spectrometer

JAM: Joypur, Ardevok and Moynar

JICA: Japan International Cooperation Agency

K: Coefficient of hydraulic conductivity

K_h: Coefficient of horizontal hydraulic conductivity

K_v: Coefficient of vertical hydraulic conductivity

KIT: Karlsruher Institut für Technologie (Karlsruhe Institute of Technology)

ld: Limit of detection

lq: Limit of quantification

LGM: Last glacial maximum

LMWL: Local meteoric waterline

log K_d value: Logarithmic distribution coefficient

MMAA: Monomethylarsonic acid

MS: Mass spectrometer

µS: Micro synchrotron

n: Number of samples

nd: Not determined

OM: Organic matter

p.a.: Per analysis

$p(CO_2)$: CO_2 partial pressure

POM: Particulate organic matter

PRM: Pre-monsoon season

PSM: Post-monsoon season

PVC: Polyvinyl chloride

r_{x-y}: Pearson's correlation coefficient between parameter X and Y

R^2: Coefficient of determination

RI: Relative intensity

rpm: Rotations per minute

SEP: Sequential extraction procedure

SI: Saturation index

SOM: Sedimentary organic matter

sp.: Species (singular)

spp.: Species (plural)

SRB: Sulphate reducing bacteria

σ: Standard deviation

TA: Total alkalinity

TC: Total carbon

TEA: Terminal electron acceptor

TIC: Total inorganic carbon

TN: Total nitrogen

TOC: Total organic carbon

TPC: Total plate count

TS: Total sulphur

USGS: United States Geological Survey

VPDB: Vienna Pee Dee Belemnite

VSMOW: Vienna Standard Mean Ocean Water

v/v: Volume to volume

WHO: World Health Organization

wt.%: weight %

XAFS: X-ray adsorption fine structure

XRD: X-ray diffraction

%-Q.: %-Quartile

%RSD: Relative standard deviation

1. INTRODUCTION

1.1   BACKGROUND

Providing freshwater for domestic use and new agricultural practices is becoming more and more important due to an increasing exploitation and pollution of the strictly limited water resources (UNESCO 2009). This is the current situation in the Bengal Delta Plain (BDP) that covers huge parts of the Indian state West Bengal and the bigger part of Bangladesh. Here, naturally occurring arsenic-bearing groundwater threatens the health of millions of residents (JOHNSTON et al. 2011, MATSCHULLAT 2000). For more than three decades, inhabitants are exposed to arsenic-enriched groundwater since millions of tube wells have been installed with support of international developing aid (SMITH et al. 2000). After symptoms of chronic arsenic (As) exposure had been diagnosed in the local population, inter-national research began in the early nineties to monitor scale and origin of this calamity (e.g., BGS & DPHE 2001). Results soon revealed that As in groundwater is inorganic and geogenic in nature, with anaerobic microbes as key players in the biogeochemical cycling of As (ISLAM et al. 2004). Development of applicable mitigation strategies developed soon as an important scope of As research, while more and more affected countries and areas were identified in Asia as well as in other regions all over the world (NRIAGU et al. 2007).

This present thesis arose from a German-Indian research project funded by the German Research Foundation (DFG) and the German Federal Ministry for Economic Cooperation and Development (BMZ), entitled "'The role of microbiogeochemical processes in releasing As from aquifer sediments in the Bengal Delta Plain (BDP): An experimental approach" (Stu 169/37-1). It is the underlying working hypothesis that As release originates from microbially mediated decomposition of organic matter (OM) under specific redox-conditions, and concurrently interferes with abiotic hydro-logical and geochemical processes on a local to regional scale.

1.2   OBJECTIVE

Stated aim of this thesis is to shed light on the biotic and abiotic controls of mobilisation and distribution of As in shallow groundwater of the BDP. The scientific focus is defined by the identification and description of two capable study sites, the characterisation of sediment samples, the conduction and interpretation of two in-situ field experiments and the interpretation of hydrochemical monitoring data from both sites. The underlying research concept is illustrated in Figure 1.1.

All results are combined in an effort to develop an advanced concept explaining the distribution of As in local groundwater, which is transferable to the entire BDP as well as other arsenic-affected regions in Asia.

The key questions are:

- How is As release linked to the availability of OM in the aquifer?
- Which biogeochemical reactions are induced by indigenous microbial communities that control the mobilisation of As into groundwater of the BDP?
- What are the specific local conditions that determine As mobilisation, and how do such conditions arise?
- How does dissolved As response to interfering abiotic processes like temporal changes in the groundwater level?

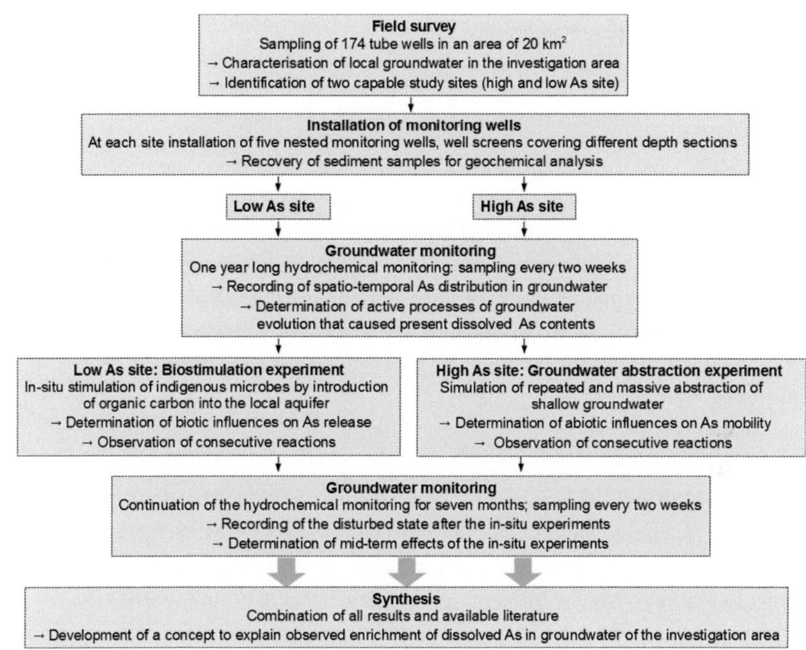

Figure 1.1: Illustration of the research concept.

1.3   STATEMENT OF NOVELTY

New aspects and approaches covered by this study are:

1. The in-situ biostimulation experiment that was conducted at the low As site to reveal biotic influences on the mobility of As. In contrast to the few previously conducted experiments using a similar approach (HARVEY et al. 2002, SAUNDERS et al. 2008), no additional TEA substances like NO_3^- or SO_4^{2-} were added.

2. Simulation of excessive groundwater abstraction at the high As study site with an accompanying hydrochemical monitoring was used to identify subsequent abiotic effects on the distribution of dissolved As in shallow groundwater. This approach is a novelty in the BDP and the results constitute a valuable contribution to the discussion of pumping impacts.

3. The detailed geochemical characterisation of sediments from the study sites provides information about potential water-sediment interactions within the surface near groundwater fluctuation zone of the BDP. Without further characterisation, this zone was previously interpreted as a potential source of As (HARVEY et al. 2006). In this study, sediments from this zone are analysed in detail to prove this hypothesis.

4. All results are combined to develop an advanced concept that explains the underlying processes of As mobilisation and spatiotemporal distribution in the BDP.

2. THE FATE OF ARSENIC IN THE ENVIRONMENT

2.1   ARSENIC IN THE ENVIRONMENT

2.1.1   GEOCHEMISTRY

Arsenic is a metalloid belonging to the nitrogen group with the atomic number 33, a molecular weight of 74.9 g mol^{-1} and ^{75}As as the only naturally occurring isotope (DE LAETER et al. 2003). With five valence electrons, it has four possible oxidation states: As(-III) (arsenide), As(0) (elemental arsenic), As(III) (arsenite) and As(V) (arsenate). Arsenic can be incorporated as a major constituent in the crystal lattice of more than 200 minerals, including elemental As(0), arsenides, sulphides, oxides, arsenates and arsenites (SMEDLEY & KINNIBURGH 2002). Common minerals that are often associated with ores are arsenopyrite (FeAsS), realgar (As$_4$S$_4$) and orpiment (As$_2$S$_3$). These minerals are generally rare in the environment, but As often occurs as trace element, especially in sulphides.

Average As contents of the upper crust range from 1.5 to 2.0 mg kg^{-1}, while average contents in rocks reach 0.5 to 2.5 mg kg^{-1}. Elevated contents appear in sedimentary iron formations and iron-rich sediments (up to 2,900 mg kg^{-1}), sandstones (0.6-120 mg kg^{-1}), coals (0.3–35,000 mg kg^{-1}), bituminous shale (100–900 mg kg^{-1}) and marine shale and mudstone (up to 490 mg kg^{-1}) (MATSCHULLAT 2000, SMEDLEY & KINNIBURGH 2001).

2.1.3 *HYDROCHEMISTRY*

Typical As concentrations in fresh water range between 1 to 2 $\mu g\ L^{-1}$ (HINDMARSH & MC CURDY 1986). In groundwater, As concentrations can increase up to several milligrams per litre, depending on parental rock contents and prevailing geohydrochemical conditions. The speciation of As in surface or groundwater is primarily determined by prevailing redox state (E_H) and pH. In oxic systems, the thermodynamically favoured form is in-organic arsenate, abbreviated in the following as As(V), which forms oxy-anions (AsO_4^{3-}) similar to phosphate (PO_4^{3-}). In acidic environments, $H_2AsO_4^-$ dominates, whilst $HAsO_4^{2-}$ appears at neutral and alkaline pH conditions (see Figure 2.1). The other two possible forms H_3AsO_4 and AsO_4^{3-} only occur under extreme acidic, respective alkaline conditions. The prevailing form in reducing environments is arsenide, abbreviated as As(III). Under acidic and neutral pH values, uncharged H_3AsO_3 dominates (SMEDLEY & KINNIBURGH 2002). Due to kinetic inhibition or microbially induced catalytic reactions, As(V) can also exist in As(III) dominated systems and vice versa (INSKEEP et al. 2002, NRIAGU et al. 2007).

The predominating forms of dissolved As in dependence of pH and E_H can be summarised in pH-E_H-stability diagrams based on thermodynamic calculations (STUMM & MORGAN 1996, Figure 2.1). When the redox potential changes from anaerobic to aerobic (for example by irrigation with reduced groundwater), As(III) reacts with atmospheric oxygen and oxidises within hours to As(V) (NEIDHARDT et al. 2012a). If groundwater holds high concentrations of dissolved Fe(II), As(III) and As(V) will rapidly co-precipitate with Fe-(oxyhydr)oxides by aeration (BERG et al. 2006).

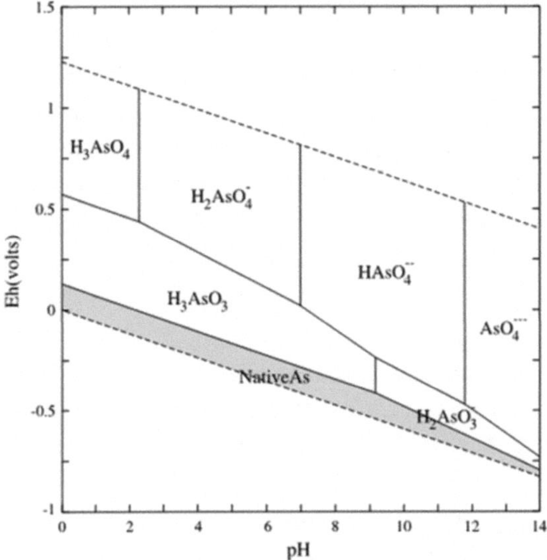

Figure 2.1: Characteristic pH-E$_H$-stability diagram for the system As-O$_2$-H$_2$O at 25°C and 0.1 MPa based on latest thermodynamic data (LU & ZHU 2011). Also included is the solid phase of As (grey shaded). The activity of As was set as 10^{-6} M (74.9 µg L^{-1}). Figure from LU & ZHU (2011), reprinted with permission from Springer through the Copyright Clearance Center.

2.1.3 BEHAVIOUR OF INORGANIC ARSENIC

Identification of the aqueous-solid-phase-interactions between aquifer sediments and the surrounding groundwater are the key to predict As mobility in aquifer systems. These interactions are strongly influenced by the activity of autochthonous microbes that interfere with the hydrochemical composition of groundwater and the mineralogical inventory of the aquifer sediments. According to the prevailing conditions, sediments can act both, as source and as sink for As. Important processes are complexation, redox-reactions and competitive anion exchange, whereas redox state and pH

7

have the most important influence on the mobility of As. Surface adsorbed As, which is either weakly adsorbed (electrostatic attraction) or strongly bound (ligand exchange), is easily accessible to interactions with dissolved compounds and microbes. Both, As(III) and As(V), have high binding affinities for Fe-(oxyhydr)oxides (e.g., goethite) and form strong bidentate complexes via ligand exchange (DIXIT & HERING 2003, MÜLLER et al. 2010, ONA-NGUEMA et al. 2005). Especially amorphous hydrous Fe-oxides and poorly crystalline Fe-(oxyhydr)oxides (e.g., ferrihydrite) have large surface areas, resulting in a chemical reactivity that is far out of proportion to their abundance (BORCH et al. 2010). Iron-(oxyhydr)oxides often occur in sediments as alteration products in form of partial coatings around mineral grains and act as important sinks for many trace elements including As (EICHE at al. 2010, GUO et al. 2007). To a lesser extent, As in sediments is associated with Mn- and Al-(oxyhydr)oxides, clay minerals, sulphates, calcium carbonates and organic acids (O'DAY 2006).

It is very difficult to estimate the sorption behaviour of dissolved As in a certain system, which is primarily influenced by pH and redox state of the solution and the presence of adsorbing mineral phases (DIXIT & HERING 2003, GOH & LIM 2004). Despite the complex nature of a multi-component system like natural aquifers, previous studies that examined the sorption behaviour of As used strongly simplified experimental setups (MOHAN & PITTMAN 2007). Additionally, the similar sorption behaviour of PO_4^{3-} and other anions plays an important role regarding competitive ion exchange and adsorption (POSTMA et al. 2007). In case of a tropical soil rich in Fe-(oxyhydr)oxides, As exchange potentials declined in order of $PO_4^{3-} \gg CO_3^{2-} > SO_4^{2-} \approx Cl^-$ (GOH & LIM 2005). Arsenic retention through adsorption generally depends on respective flow velocities, available binding sites and adsorption partners, as well as concentrations of competing ions and solutes in groundwater.

2.1.4 TOXICITY

Arsenic has influenced human history for thousands of years according to its extremely high toxic potential and can therefore be entitled as the king of poison. Arsenic uptake causes acute as well as chronic intoxications, even at very low doses. Acute arsenic poisoning mostly manifests after accidents with pesticides or homicidal intensions, while chronic intoxication is mainly derived from oral ingestion of arsenic-enriched drinking water (MELIKER & NRIAGU 2007). Resulting effects of chronic poisoning are complex and depend on the prevailing chemical form, whereas both inorganic species As(V) and As(III) are much more reactive than methylated organic forms (HOPENHAYN 2006, WHO 2003). Inorganic As is supposed to act genotoxic, carcinogenic and teratogenic (WHO 2003). Due to its similarity to phosphate, As(V) can interact with up to 200 enzymes, most of them being part of the adenosine-tri-phosphate (ATP) synthesis pathway or the DNA synthesis and repair system (ABERNATHY et al. 1999, ISLAM 2008). Reduced inorganic As(III) is considered even more toxic to human organism, which results from its high affinity for reactive thiol groups of enzymes (KNOWLES & BENSON 1983). The principal organ of As metabolism is the liver, where inorganic As is methylated to dimethylarsinic acid (DMAA) and monomethylarsonic acid (MMAA), before it is excreted via urine. The half-life of inorganic As compounds in the human body is 2-40 days after resorption, but a continuously uptake results in enduring enrichment in liver, kidneys, heart, lungs and ectodermic tissues (POMROY et al. 1980). Chronic exposure to increased concentrations of inorganic As, especially As(III), is known to entail severe diseases and its carcinogenic character promotes an increased appearance of cancer (skin, lung, bladder and liver) in affected populations, which was observed in various case studies in Bangladesh, Taiwan and China (e.g., CHEN et al. 1992, KAPAJ et al. 2006, SMITH et al. 2000). Another characteristic expression is arsenicosis, a collective term for skin lesions like keratosis, hyperkeratosis and pigmentary abnormalities of the extremities (AHMAD et al. 1997).

One of the most famous examples for an endemic occurrence of arsenicosis is an area in the southwest of Taiwan, where local villagers had changed their drinking water source from surface water to arsenic-enriched artesian groundwater in the 1920's. This undiscovered exposure soon caused symptoms of chronic As intoxications, which were first described in the 1950's as "black foot disease" (IPCS 2001). Long-term cohort studies in affected villages of Vietnam, China and Bangladesh imply that chronic As uptake may also trigger foetal loss and infant death, development of diabetes mellitus, cardio-vascular disease and eventually neurotoxic effects and inhibition of children's mental development (ALAM et al. 2002, ARGOS et al. 2010, FUJINO et al. 2006, LIN et al. 2004, RAHMAN et al. 1998 & 2007, WASSERMANN et al. 2004). A profound overview over health effects related to As can be found in the work of NRIAGU (1994). To the present day, no proper therapy for arsenicosis exists, which is why mitigation strategies are the only available possibility to avoid diseases related to chronic As uptake. It is further very problematic to assess a proper threshold value for drinking water and food, since no dose-response relationship exists according to the carcinogenic character of inorganic As. The WHO released a provisional guideline value for total As in drinking water of 10 µg L^{-1}, based on the level that can be achieved through practical treatment methods (WHO 2011). In India, there is actually a legal limit of 50 µg L^{-1} for the total As concentration in drinking water in force (Indian Standard Specifications for drinking water IS 10500, reaffirmed 1993).

2.1.5 *WORLDWIDE OCCURRENCE AND THE HUMAN HEALTH DISASTER IN THE BENGAL DELTA PLAIN*

Concentrations of As in natural aquifers exceeding 10 µg L^{-1} were reported from all over the world, e.g., Argentina, Australia, Bolivia, Cambodia, Chile, Ecuador, El Salvador, Honduras, Hungary, Mexico, Nepal, New Zealand, Nicaragua, Myanmar, Philippines, Taiwan, Thailand, Uruguay and the United States. In total, a population of more than 100 million people relies in these affected areas on groundwater as often only available drinking water source (NRIAGU et al. 2007). The worldwide largest affected areas are located in deltaic floodplains of Asia, comprising the BDP in India and Bangladesh and the Red River Delta in Vietnam (BERG et al. 2001, VAN GEEN et al. 2006). Today, still new incidences are discovered in South America and only little is known about the situation on the African continent.

The situation in West Bengal and Bangladesh won notoriety due to the epidemic spread of arsenicosis two decades after millions of tube wells were installed with the help of international development aid in the 1970's (CHARLET & POLYA 2006, POLYA & CHARLET 2009). Pathogenic polluted surface water was replaced by supposedly safe groundwater as principal drinking water source for the rural population, but the water was not tested for As. Partly extremely As(III)-enriched groundwater provoked a creeping mass poisoning, which was dubbed by the WHO as the largest mass poisoning of a population in human history (SMITH et al. 2000). The common use of groundwater for irrigation purposes raises additionally the risk of As entering the food chain. Rice, the fundamental crop grown in the BDP, does fortunately not enrich As in its grains, since rice roots are naturally covered with a coating of Fe-(oxyhydr)oxides that serve as an effective barrier (NORRA et al. 2005). Other crops seem to efficiently prevent As enrichment in the grains, although other plant tissues like roots or leaves may accumulate As (NEIDHADRT et al. 2012a).

2.1.6 *MITIGATION STRATEGIES*

The key problem of rural areas in underprivileged countries is the lack of a centralised water supply, where As can be effectively removed during the step of Fe-removal via groundwater aeration. A large-scale treatment of affected aquifers is impossible due to the immense dimensions and the diffuse spreading of dissolved As. In Vietnam, people successfully use self-made sand filters to remove undesired high concentrations of dissolved Fe. During filtration, Fe-(oxyhydr)oxides precipitate because of aeration, additionally co-precipitating dissolved As(III/V). BERG et al. (2006) demonstrated that this technique removed an average of 80 % of the initial total As concentrations that ranged from 10 to 382 µg L^{-1}. The challenge is to recognise when the filter material needs to be exchanged, and to dump the loaded filter sands safely under oxic conditions in order to prevent re-mobilisation of As. The same applies to more sophisticated filter systems, which are provided in form of public central water supply wells to local villagers. In West Bengal and Bangladesh, the socio-economic situation of the plain's inhabitants often hinders a successfully implementation and maintenance of commercial filter techniques to the present day. Pond sand filters and rain water harvesting are alternative options to obtain drinking water, but inadequate maintenance often bears the risk of contamination with pathogenic microorganisms (AHUJA 2008). A promising approach is the identification of safe wells by using cheap and simple As measuring methods like field test kits (KINNIBURGH & SMEDLEY 2001, WORLD BANK 2005, YU et al. 2003). In recent years, the local government favoured the installation of deep tube wells (>150 m depth) to maintain arsenic-free water to the public. Installation of centralised water treatment facilities enables meanwhile at least wealthy families in larger villages access to treated water. In smaller villages, households who can afford it try to avoid arsenic-rich groundwater by installation of deeper private tube wells (CHARLET et al. 2007).

A new and innovative approach to predict endangered areas is the use of modelling software and easy available surface data (topography, spatial data) (WINKEL et al. 2008). The Japan International Cooperation Agency (JICA) concluded that the most appropriate As mitigation strategy in low income countries is an enduring area-wide information and education campaign that imparts the development of a public awareness and knowledge of the correct use of simple filter systems (JICA 2003).

2.2 MOBILITY OF ARSENIC IN AQUIFER SYSTEMS

2.2.1 BACKGROUND

In order to understand local distribution patterns of As in an affected area, it is mandatory to reveal the underlying mechanisms of release and transport. In general, As can be released into groundwaters either by natural processes or by anthropogenic activities. Natural processes involve mainly chemical and biological weathering of rocks and sediments, hot springs and thermal waters, volcanic eruptions and fumaroles. About 80 % of arsenic utilised by human is released diffusely into the environment, for example as constituent of herbicides, insecticides, desiccants, defoliants, feed additives, wood preservation agents, pigments, drugs and alloying elements (NRIAGU et al. 2007). Industrial activities, especially mining, smelting and coal combustion, are further important point sources releasing As through mine drainage, waste water and exhausted air (IPCS 2001).

In case of the BDP, geogenic As is naturally released from saturated aquifer sediments under anaerobic conditions by water-rock interactions, which are described in the following. Arsenic was found to be mainly associated with Fe-(oxyhydr)oxides and sulphides, which are believed to originate from the Himalaya Mountains (AKAI et al. 2004). Here, As is released from minerals (e.g., realgar, orpiment, arsenian pyrite) via physical weathering of igneous and highly metamorphic rocks under oxic conditions. After release, As immediately co-precipitates and/or adsorbs onto Fe-(oxyhydr)oxides that precipitate in form of nanoparticles, colloids and coatings (RAISWELL 2011, HASSELHÖV & KAMMER 2008, EICHE et al. 2010). In the following, the three huge streams of the BDP (Ganges, Brahmaputra and Hooghly River) and their tributaries transport a mixture of arsenic-enriched primary mineral fragments and secondary arsenic-hosting Fe-particles as suspended matter into the BDP. With decreasing relief energy, sediments get buried together with high amounts of OM in the huge, coastal near floodplains. Microbial decomposition of buried organic carbon soon generates reducing conditions in the subsurface. Attempts to directly link high As concentrations within Holocene aquifers to the respective mineralogy have failed, and it remains unclear whether human activities additionally affect As release and distribution (BGS & DPHE 2001). The ultimate mechanism that drives As enrichment in groundwater of the BDP is still object of intense debate as described in the following.

2.2.2 *OXIDATION HYPOTHESIS*

One of the first proposed mobilisation mechanisms was the oxidation hypothesis (DAS et al. 1996). Arsenic is supposed to be released via oxidation of arsenic-enriched sulphides within the partly oxic groundwater fluctuation zone, followed by downward transport during surface recharge. The underlying assumption is that high concentrations of dissolved As require high sedimentary As contents. In early studies, arsenic-enriched pyrite was found and therefore designated as origin of dissolved As in groundwater (CHOWDHURY et al. 1999, DAS et al. 1996). Chemical weathering of pyrite-rich barren rock deposits in mining areas is known to have caused large-scale As releases via acid mine drainage, e.g. in South America (BUNDSCHUH et al. 2010, MORIN & CALAS 2006). HARVEY et al. (2002) further argue that a widespread and excessive groundwater with-drawal in many areas of the BDP causes enduring decreases in the local water levels and therefore aeration of normally saturated sediments, followed by additional release of As. This publication provoked an open clash of opinions settled in renowned journals since arsenic-enriched groundwater was also reported from large-scale areas in the BDP, which are considered unaffected by groundwater extraction (AGGARWAL et al. 2003, HARVEY et al. 2003, VAN GEEN et al. 2003). Another point of criticism is the fact that high contents of dissolved As do not necessarily require high sedimentary As contents. Excluding flow and influences of transport and re-adsorption, the release of one mg kg^{-1} As in one m^3 of sediments will result in a concentration increase of 2,650 µg L^{-1} in ground-water, assuming a pore volume of 50 % and a sediment density of 2.65 g cm^{-3}.

Hence, prevailing conditions that cause release of comparatively small amounts of available sedimentary As and a slow groundwater flow are under certain conditions more important than absolute contents present in the sediments.

2.2.3 IRON REDUCTION HYPOTHESIS

In the past few years, indications grew stronger that mobility, reactivity, bioavailability and toxicity of As are primarily determined by biogeochemical transformations. Positive correlations between As and Fe(II) contents in groundwater point at a potential influence of redox reactions, which cause reductive dissolution of Fe-minerals (MC ARTHUR et al. 2001, NICKSON et al. 1998 & 2000, RAVENSCROFT et al. 2001, STÜBEN et al. 2003, WAGNER et al. 2005). Numerous field studies concluded that As is released passively through microbial oxidation of OM in absence of O_2 (ISLAM et al. 2004, MC ARTHUR et al. 2004). Nowadays, there is broad agreement within the scientific community that extensive reductive dissolution of arsenic-hosting Fe-(oxyhydr)oxides is the key factor of As release in Asia (FENDORF et al. 2010, SAUNDERS et al. 2008). The gist of this hypothesis dates back to studies of DEUEL & SWOBODA (1972), GULENS et al. (1979) and the pioneering work of LOVLEY and co-workers in the field of Geomicrobiology during the last three decades (e.g., LOVLEY et al. 1986). This hypothesis extensively was tested in the following and was further supported by laboratory column, batch and microcosm experiments (e.g., FAROOQ et al. 2010, ISLAM et al. 2004, RADLOFF et al. 2007, VAN GEEN et al. 2004).

Microbiologically mediated redox reactions related to anaerobic dissimilatory (energy yielding) degradation of OM in aquifer systems proceed with a successive consumption of terminal electron acceptors (TEA) and a decreasing energy yield, until all available degradable OM (solid and dissolved) is exhausted (BORCH et al. 2010, CHAPELLE 1993, JURGENS et al. 2009, LANGMUIR 1997, MC MAHON & CHAPELLE 2008). In a static pool of water, increasing depth is equivalent to increasing age and therefore reaction time, allowing the establishment of horizontally layered redox zones. Each zone is dominated by one specific TEA consuming process and the respective metabolic pathway used by involved microorganisms (Figure 2.2 and Table 2.1).

After depletion of dissolved O_2, the system turns from oxic to anoxic, accompanied by a continuously declining redox potential (CHRISTENSEN et al. 2000, STUMM & MORGAN 1996). This sequence is the result of a dynamic competition between different heterotrophic microbes and the metabolic pathways they are able to use for decomposition of OM. Those species are preferred that can gain the highest energy yield from available OM and TEA in a certain environment (ACHTNICH et al. 1995).

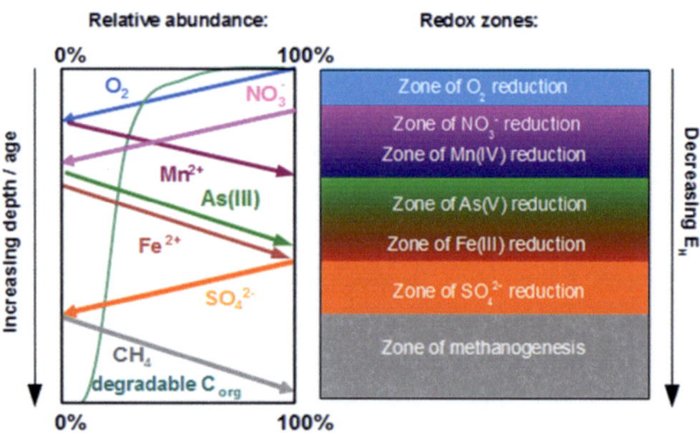

Figure 2.2: Microbial consumption of available and degradable organic carbon (C_{org}) in a static pool of water, causing sequential depletion of available TEA and a decrease of the redox potential with depth and age. Different redox reactions may overlap and occur parallel to each other, such as NO_3^- and Mn(IV) reduction. Figure based on BORCH et al. (2010), KOCAR & FENDORF (2009), STUMM & MORGAN (1996).

When iron reduction arises, Fe-(oxyhydr)oxides get dissolved, causing a passive release of adsorbed and incorporated trace elements including As (LOVLEY 1995 & 1997). Amorphous (hydrous ferric oxides) and poorly crystalline (e.g., ferrihydrite) Fe-phases are the favoured source of Fe(III) for microbes, although crystalline and thermodynamically more stable minerals like lepidocrocite, goethite, hematite and magnetite can be utilised, too, when the redox potential declines accordingly (LOVLEY & PHILLIPS 1986, LOVLEY & PHILLIPS 1987).

Table 2.1: Ecological redox sequence in order of decreasing energy gain (after STUMM & MORGAN 1996).

TEA consuming process	TEA	Redox-reactions
Aerobic respiration	O_2	$O_2 + \{CH_2O\} \rightarrow H_2O + CO_2$
Nitrate reduction	NO_3^-	$NO_3^- + 2\{CH_2O\} + 2H^+ \rightarrow NH_4^+ + 2CO_2 + H_2O$
Manganese reduction	$Mn(IV)$	$4MnO_2 + 2\{CH_2O\} + 8H^+ \rightarrow 4Mn^{2+} + 2CO_2 + 6H_2O$
Iron reduction	$Fe(III)$	$4Fe(OOH) + \{CH_2O\} + 8H^+ \rightarrow 4Fe^{2+} + CO_2 + 7H_2O$
Sulphate reduction	SO_4^{2-}	$SO_4^{2-} + 2\{CH_2O\} + H^+ \rightarrow HS^- + 2CO_2 + 2H_2O$
Methanogenesis	$\{CH_2O\}$	$2\{CH_2O\} \rightarrow CH_4 + CO_2$

It is mandatory to understand that the biogeochemical cycling of an element like As or Fe never depends on the activity of a single microbial species (COZZARELLI & WEISS 2007, LOVLEY & PHILLIPS 1989). Microbes form species-independent, connected networks (consortiums) that are adapted to the present environment (quality of OM, prevailing Fe-minerals, pH, temperature, intra-species interactions, etc.) enable their members to degrade OM stepwise (CUMMINGS & MAGNUSON 2007, KONHAUSER 2007). Involved organism mainly belong to the domain of *Bacteria* and therein to the heterotrophic dissimilatory Fe(III)-reducing bacteria (FeRB), which couple the reduction of Fe(III) with the oxidation of metabolic products released by synergistic bacteria (alcohols, short- and long-chain fatty acids, mono-aromatic compounds or H_2) (KONHAUSER et al. 2011, LOVLEY & PHILLIPS 1989, LOVLEY et al. 2004). The most important fermentation product for microbial Fe(III) reduction in soils and sediments is acetate (CH_3OOH), as demonstrated by *Geobacter metallireducens*, an ubiquitous species in aquifer environments:

$$2CH_3OO^- + 8Fe(III) + 2H_2O \rightarrow 2HCO_3^- + 8Fe(II) + 8H^+ \qquad (2.1)$$

(LOVLEY 1993)

In addition to the release of Fe(II), this metabolic reaction causes formation and release of protons and bicarbonate, which trigger in turn subsequent geochemical reactions like carbonate dissolution.

Depending on the availability of TEA and OM, microbial communities and resulting geomicrobiological reactions evolve over time. Importantly, Fe(III) reduction may even occur when other redox reactions are thermodynamically favoured (RODEN 2003) and Fe(III)-reducing microbes are often able to use other metabolic pathways as well, like the reduction of Mn(IV) (LOVLEY 1993). A complete reduction of the Fe(III) pool in natural aquifer systems is rather unlikely, since amounts of available OM are commonly limited compared to ubiquitously occurring Fe(III)-minerals (BORCH et al. 2010).

At circum-neutral pH values, available dissolved Fe(III) is quite rare since Fe(III)-minerals are poorly soluble under such conditions. To overcome the resulting lack, many FeRB excrete specific low molecular weight organic compounds that serve as Fe(III)-electron shuttles (WOLF et al. 2009). For example, siderophores specifically adsorb onto Fe(III)-bearing mineral surfaces, thereby mobilising Fe(III) via surface complexion (KONHAUSER et al. 2011, RODEN et al. 2010). Most FeRB are able to release different kinds of electron shuttles, depending on the present mineral phases (PAGE & HUYER 1984). Despite this, other species are able to directly mobilise and uptake Fe(III) after adsorption to the mineral surface via pili and flagella (WOLF et al. 2009).

It is not easy to proof that Fe(III) reduction is a currently active redox process in an aquifer system. Measuring of the redox potential is not sufficient to exclude Fe(III) reduction since the respective pathway may also be active in redox zones beyond the characteristic redox potentials. Determination of genes necessary for Fe(III) reduction via molecular biological methods is also a non-reliable predictor, because a proof of presence is no evidence for respective metabolic activity (CUMMINGS et al. 2000, CUMMINGS & MAGNUSON 2007).

When Fe(III) reduction is the predominating redox process, the following conditions require to be fulfilled:

- Absence of O_2 and NO_3^- as competing TEA;

- Presence of Fe(II) (dissolved and/or in solid form);

- Degradation of acetate to HCO_3^-;

- Absence of high molybdate concentrations that would inhibit microbial oxidation of acetate;

- High concentrations of SO_4^{2-} and concomitant low S^{2-} contents (CUMMINGS & MAGNUSON 2007, LOVLEY et al. 1994).

In addition, the occurrence of magnetite is considered a circumstantial evidence for past Fe(III) reduction (see chapter 2.2.4, mineral transformation processes) (CUMMINGS & MAGNUSON 2007, GIBBS-EGGAR et al. 1999).

Similar processes are associated with microbial reduction of Mn-oxides. Microbes capable of Mn(IV) reduction are also ubiquitous in sediments as well as Mn-oxides, which may act as arsenic-hosting mineral phases, too (LOVLEY et al. 1993, 1995, 2004).

2.2.4 DISSIMILATORY ARSENATE RESPIRING PROKARYOTES

Some microorganisms are able to reduce As(V) to As(III) via respiration pathways (dissimilatory arsenate respiring prokaryotes, DARPs). The mobility of As largely depends on the prevailing pH and redox potential and on the availability of potential sorption phases, especially Mn-oxides and Fe-(oxyhydr)oxides (LEAR et al. 2007, OREMLAND & STOLZ 2005). In anoxic environments with near-neutral pH values, As mobility is believed to increase after reduction to As(III) (HARVEY et al. 2002, KINNIBURGH & SMEDLEY 2001). Despite this, laboratory experiments have demonstrated that this statement is an oversimplification and that an estimation of the species dependent As mobility is very difficult in natural aquifer systems (HERING & KNEEBONE 2002). Sorption behaviour of dissolved As(III) and As(V) strongly depends on the solid-to-solution ratio of both As species; the quantity and type of adsorbing mineral phases, and the concentration of phosphate, which is known to act as a strong competitor for As binding sites (MOHAN & PITTMAN 2007).

Following thermodynamic analysis, As(V) reduction appears in range of similar redox potentials like Fe(III) reduction under most environmental conditions (KOCAR & FENDORF 2009). The presence of DARPs can be validated by screening for the referring arsenate respiratory reductase gene (*arrA*), which is necessary for the formation of As(V) reductase (MALASARN et al. 2004). A presence of the *arrA*-gene in local microbial communities means neither that bacteria actually synthesise arsenate reductase, nor that it is an essential metabolic pathway, since many microbes have different possible respiratory mechanisms. For example, the *arrA*-gene could have been identified in many FeRB (CAMPBELL 2006, ZOBRIST et al. 2000). Additionally, some bacteria detoxify As(V) by reduction, using the so called arsC pathway (CAMPBELL et al. 2006). In groundwater of the BDP, As(III) is often the predominating species of dissolved As and it is most likely that both, Fe(III) and As(V) reduction, are closely linked (KAPPLER 2011, LLOYD & OREMLAND 2006).

X-ray absorption near edge structure (XANES) measurements support the assumption that a direct reduction of surface adsorbed As(V) may play an important role in As mobilisation during the early sedimentary diagenesis, while in matured sediments As(III) dominates (ROWLAND et al. 2005).

2.2.5 *THE ROLE OF ORGANIC MATTER*

The variety of microbially mediated degradation processes involved in mobilisation of As emphasizes the relevance of OM. Pristine aquifers are commonly carbon limited systems (COZZARELLI & WEISS 2007), but the high sedimentation rate in active deltas and floodplains enables a rapid burial of considerable amounts of organic matter (KINNIBURGH & SMEDLEY 2001, QUICKSALL et al. 2008). The lack of O_2 and light generally impedes biodegradation in saturated aquifers, causing reaction rates to slow down. Labile, reactive or easily degradable are partly synonymous adjectives to describe organic carbon (e.g., sugars, amylum) that can be readily degraded by a wide range of microbes including FeRB, and therefore decreases in aquifer sediments with time and depth. In mature sediments, only recalcitrant organic matter like plant fibres and long-chain molecules carrying double bonds and aromatic compounds (e.g., humic and fulvic acids, petroleum products) remain preserved (ROWLAND et al. 2006 & 2007). Reduced humic substances function as important electron shuttles and dominate the dissolved organic carbon (DOC) fraction in Fe(III)-reducing aquifer parts (KAPPLER et al. 2004, LOVLEY et al. 1996, MLADENOV et al. 2010). Hence, DOC plays a dual role depending on the degradability: labile DOC serves as electron donor in microbial respiration, while persistent humic substances are used as electron shuttles during Fe(III) reduction. Dissolved organic molecules are mobile and can be easily transported along hydrological flow paths. In deltaic regions of Asia, wetlands, ponds and fields may serve as important potential sources of infiltrating OM, especially during monsoon rains, when the surface recharge drastically increases (KOCAR et al. 2008).

Surface-derived OM is furthermore one of the key issues in the discussion of anthropogenic influences on the release of As (2.2.7).

2.2.6 *FURTHER MICROBIAL INTERACTIONS*

The following processes and mechanisms have the potential to effect As release in natural environments, but are considered to play minor roles in context of the processes involved in aquifers of the BDP.

Toxic effects arise from the high similarity between As(V) and PO_4^{3-} that can also affect microbial respiration. Hence, some bacteria are capable to reduce dissolved As(V) to As(III) or even further to methylated organic As compounds in order to detoxify it. Experiments with *Shewanella sp.* conducted by CAMPBELL et al. (2006) revealed that the formation of ArsC reductase (an enzyme catalysing As(V) reduction) triggers at a threshold concentration of 100 µmol L^{-1} As(V). Such a relatively high threshold concentration indicates that this process is most likely not relevant in shallow Asian aquifers, where As(III) is the predominating species in groundwater and even highest total As concentrations clearly remain below 100 µmol L^{-1} (SMEDLEY & KINNIBURGH 2002). Most important organic As compounds are DMAA, MMAA and volatile arsine (AsH_3). Methylation of As is considered to primarily occur in marine and geothermal environments, but not in low-temperature aquifers (BEDNAR et al. 2004, LE 2002).

In contrast, arsenic-oxidizing bacteria are capable to oxidize As(III) to As(V) in order to yield metabolic energy (heterotrophic arsenide oxidising bacteria, HAO) (OREMLAND & STOLZ 2005, SALMASSI et al. 2002). Chemolitho-autotrophic bacteria (CAO) couple the oxidation of As(III) with the reduction of O_2 or NO_3^- to gain energy, a process that does not depend on organic matter as electron donor (JACKSON et al. 2001, SENN & HEMOND 2002).

2.2.7 INFLUENCES OF HYDROLOGY AND ANTHROPOGENIC ACTIVITIES

The fate of aqueous As is further determined by the local hydrogeology, which directs the transport of arsenic-enriched groundwater within the aquifer. In addition, the reaction time for groundwater-sediment interactions is set by the flow velocity and becomes an important factor in case of kinetically controlled hydrochemical balance reactions, like competitive surface exchange, biotransformation processes and dissolution of mineral phases. Reactions following hydrochemical water-rock interactions can be abiotic and/or biotic and influence As mobility by either inducing additional release or retention. STUTE et al. (2007) postulated a significant positive correlation between groundwater age and As concentrations in shallow groundwater of the BDP. They observed a nearly constant increase in As over time until a certain maximum was reached, and assumed that As release is the result of a kinetically inhibited balance reaction. However, this hypothesis cannot explain the typically bell-shaped depth distribution of As and the fact that older aquifer parts are practically arsenic-free.

The monsoon climate strongly affects the regional hydrology in entire Asia, especially in the BDP. Seasonal changes between the dry season and the annual monsoon rain induce here pronounced oscillations in the local water table, reaching annual net fluctuations of up to 5 m (HARVEY et al. 2005). Accompanying shifts in the redox potential within affected surface near aquifer sediments comprise oxic conditions during the dry season as well as reducing milieus following monsoonal recharge.

Groundwater flow within the Bengal Basin follows a superior regional flow direction from the Himalayan Mountains in the north, to the Bay of Bengal in the south. The horizontal hydraulic conductivity is normally higher than in vertical direction, thereby preserving heterogeneous vertical As distributions once they have been established (MICHAEL & VOSS 2009a). In addition, the generally flat landscape results in extremely low ground-water flow velocities and long-lasting residence times, inducing a subsequent enrichment of As when it is locally released.

As long as permanently new As is released, a flush-out into the Bay of Bengal either by groundwater (dissolved As) or surface streams (part of suspended matter) is limited.

Extreme low flow velocities further provoke a high vulnerability towards pumping (MICHAEL & VOSS 2009b). For example, groundwater abstraction for irrigation purposes was propagated in time of India's green revolution to overcome the annual dry season between October to April. During the past four decades, about 10 million irrigation and tap water wells had been installed in the Bengal Basin, providing drinking water to more than 100 million people and affecting the hydrology at a regional scale (BHATTARACHARYA et al. 2004, HARVEY et al. 2005). This has caused partly massive depletions in local hydraulic heads, causing aeration and oxidation of surface near reduced sediments, which could have entailed additional As release according to the oxidation hypothesis (chapter 2.2.2). Another severe problem is the vulnerability of arsenic-free aquifers towards attraction of arsenic-rich groundwater in direction of the pumping wells. In case of a newly installed pumping field in Hanoi (Vietnam), the originally arsenic-free water soon increased in As due to attraction of arsenic-enriched groundwater (NORRMAN et al. 2008). Additionally, increasing As concentrations were recently reported from deep wells situated in the Bengal Basin as well as in the Vietnamese Red River Delta (BURGES et al. 2010, FENDORF et al. 2010, MUKHERJEE et al. 2011, NORRMAN et al. 2008, VAN GEEN 2008, WINKEL et al. 2011).

Excessive irrigation pumping may further cause infiltration of fresh and organic-rich water into mature, carbon-limited aquifer parts, what could stimulate indigenous FeRB and/or DARPs (HARVEY et al. 2002, NEUMANN et al. 2009, SUTTON et al. 2009, VAIDYANATHAN 2011). Additional OM can by derived from anthropogenic sources like agriculture, pit latrines and artificial ponds (NATH et al. 2008, POLIOZOTTO et al. 2008). Especially the potential influence of the mainly artificial ponds on As release remains unclear, as demonstrated by contradictory studies (FAROOQ et al. 2010, HARVEY et al. 2006, SENGUPTA et al. 2008, NEUMANN et al. 2009).

2.2.8 BIOTIC VERSUS ABIOTIC EFFECTS CONTROLLING ARSENIC MOBILITY

Stepwise anaerobic degradation of OM can directly affect the mobility of As in various ways (chapter 2.2.4). Despite the extensive research on microbiological controls on As mobilisation, there is no consensus regarding the potential influence of consecutive abiotic reactions. Following microbial Mn(IV)- and Fe(III)-reduction, concentrations of HCO_3^-, Mn(II) and Fe(II) increase in groundwater (see chapter 2.2.3, equation 2.1), which is often characteristic for arsenic-enriched groundwaters in Asia. Degradation of OM and reductive dissolution of Fe(III)-minerals further releases considerable amounts of PO_4^{3-} to local groundwater, which is capable to interfere with As adsorption (APPELO et al. 2002, DIXIT & HERING 2003, GOH & LIM 2004). In contrast, high amounts of dissolved Fe(II) can cause chemical reduction of surface adsorbed As(V), thereby potentially increasing its mobility (ITAI et al. 2010). Recent laboratory experiments (e.g., HANSEL et al. 2003, KOCAR et al. 2006, PEDERSEN et al. 2005, TUFANO & FENDORF 2008) have further demonstrated that dissolved Fe(II) is capable of transforming thermodynamically less stable sedimentary iron phases (e.g., ferrihydrite) rapidly into highly ordered mineral structures like magnetite (ISLAM et al. 2005, O'LOUGHLIN et al. 2010, YANINA & ROSSO 2008). The newly formed iron phases promote in turn retention of dissolved As, thereby decoupling As mobilisation from the reductive dissolution of Fe-(oxyhydr)oxides. Such transformation processes further affect the As surface adsorption capacity of affected Fe-minerals largely. For example, transformation of ferrihydrite to goethite causes a net loss of the As adsorption capacity, but re-adsorbed As appears to be stronger bound (COKER et al. 2006, HANSEL et al. 2003, PEDERSEN et al. 2005). Hence, a complex environment of interconnected biotic and abiotic processes arises in Fe(III)-reducing redox zones and induces competing As release and retention (COKER et al. 2006, HERBEL & FENDORF 2006, KOCAR et al. 2006, PEDERSEN et al. 2006).

Another example for consecutive abiotic reactions affecting As mobility is the formation of arsenic-retaining sulphides during SO_4^{2-}-reduction. Microbially released H_2S reacts with dissolved Fe(II) and precipitates, for example, as pyrite, thereby incorporating dissolved As (KIRK et al. 2010). More interestingly, some bacteria that are capable to reduce sulphate (SRB) also have the ability to reduce As(V) (NEWMAN et al. 1998). HÉRY et al. (2010) suggested a controlled introduction of labile organic carbon into SO_4^{2-}-reducing aquifers as an efficient remediation technique to remove As from groundwater with low Fe(II) concentrations.

Groundwater flow enables the connection of different redox zones within an aquifer and therefore relocation of released As, dissolved TEA (NO_3^-, SO_4^{2-}), DOC and reactive metabolic products like Fe(II), H_2S, and acetate, which trigger in turn subsequent reactions. Hence, it is nearly impossible to strictly distinguish between abiotic and biotic processes that affect the mobility of As in an aquifer, where all processes and mechanisms are inseparably connected. It is further extremely difficult to extract information about the quantitative influence of each process due to the large variety of potentially superimposing mechanisms. Laboratory experiments like column, batch or microcosm experiments are helpful to reproduce and focus on one or two mechanisms, but they can never cover the full range of naturally occurring processes. On the other hand, the evaluation of in-situ experiments is problematic as well, according to the high complexity of natural aquifer systems.

3. THE BENGAL DELTA PLAIN

3.1 GEOGRAPHY AND GEOMORPHOLOGY

The study area is located in the Indian part of the Bengal Basin and is situated in the Himalayan foreland at the junction of the Indian, Eurasian and Burmese Plates. On the west and northwest, the Bengal Basin is bounded by the Rajmahal Hills, and from northeast to east, it is flanked by the Shillong Plateau, the Tripura Hills and the Indo-Burmese Fold Belt (Figure 3.1). The southern delta and the adjacent floodplain are bounded by the Bay of Bengal, which forms the world's largest submarine fan (MUKHERJEE et al. 2009). Today's basin is the largest fluvio-deltaic sedimentary system on earth, occupying an area of about 200,000 km^2, where approximately 120 million people live in three federal states of India (Bihar, West Bengal, Assam) and the entire state of Bangladesh (ALAM et al. 2003) (Figure 3.2).

Present geomorphologic characteristics divide the Bengal Basin into Holocene lowland and Pleistocene uplands. Pleistocene units include four terraces, which are interpreted as remains of former floodplains. Two of them flank the basin in front of the Rajmahal Hills and the Tripura Hills, while the Barind and Madhupur terraces are situated inside the basin (Figure 3.1). Although outcropping Pleistocene formations are similar to Holocene units, they are characterised by a reddish brown colour, are mottled and relatively dry, have low organic contents and contain nodules of secondary iron phases and carbonates (BGS & DPHE 2001). The Holocene lowland can be subdivided into the Holocene alluvial floodplain (fluviatile, freshwater dominated floodplains) and the intergradient delta (the saline lower delta in the south), which are often aggregated as the Ganges-Brahmaputra-Meghna flood and delta plain, or simply as the Bengal Delta Plain, henceforth referred to as BDP (MUKHERJEE et al. 2009).

Four different Holocene geomorphologic main units can be distinguished in the Bengal Basin:

- The alluvial fans located at the foothills of the Himalaya, which are mainly composed of fresh coarse sand and gravel;

- The Tippera Surface near the Tripura Hills;

- The Sylhet Basin located in the northeast;

- The central floodplain and the southern delta of the BDP that cover together an area of about 10^5 km^2 and represent the central formation (MUKHERJEE et al. 2009).

In the central floodplain and the southern delta, numerous sub-deltas of the three rivers (Ganges, Brahmaputra and Meghna) overlap and create a dense system of adjacent alluvial flood- plains. The elevation of this areal reaches from 15-20 m above the sea level (asl) in the northwest, to 1-2 m asl near the southern shoreline (Figure 3.2). Meanwhile inactive Quaternary faults had changed the flow of important tributaries for several times, and created numerous interconnected abandoned tributary channels that are distributed among the entire floodplain and delta complex (ACHARYYA et al. 2000).

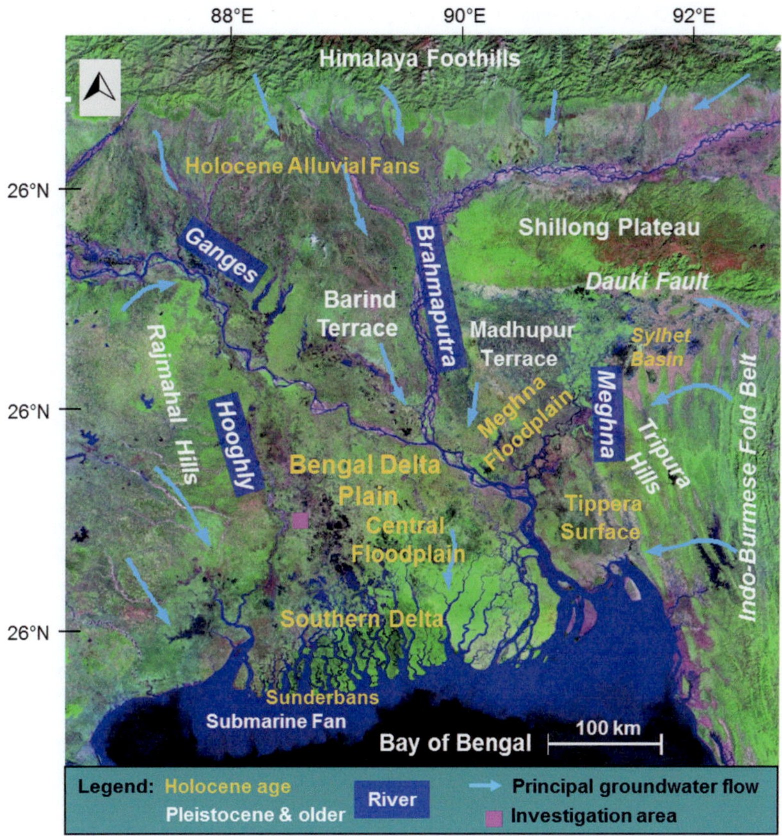

Figure 3.1: **Pseudo colour mosaic picture of the Bengal Basin created with NASA World Wind (Landsat 7 image). Coloration emphasizes the presence of water (blue), vegetation (green) and bare soil (purple). The picture is completed by landform features, location of the investigation area and principal regional flow directions of groundwater (after MICHEAL & VOSS 2009a,b and MUKHERJEE et al. 2007, 2009). Underlying pseudo colour image provided by MDA Federal Information Systems for NASA World Wind.**

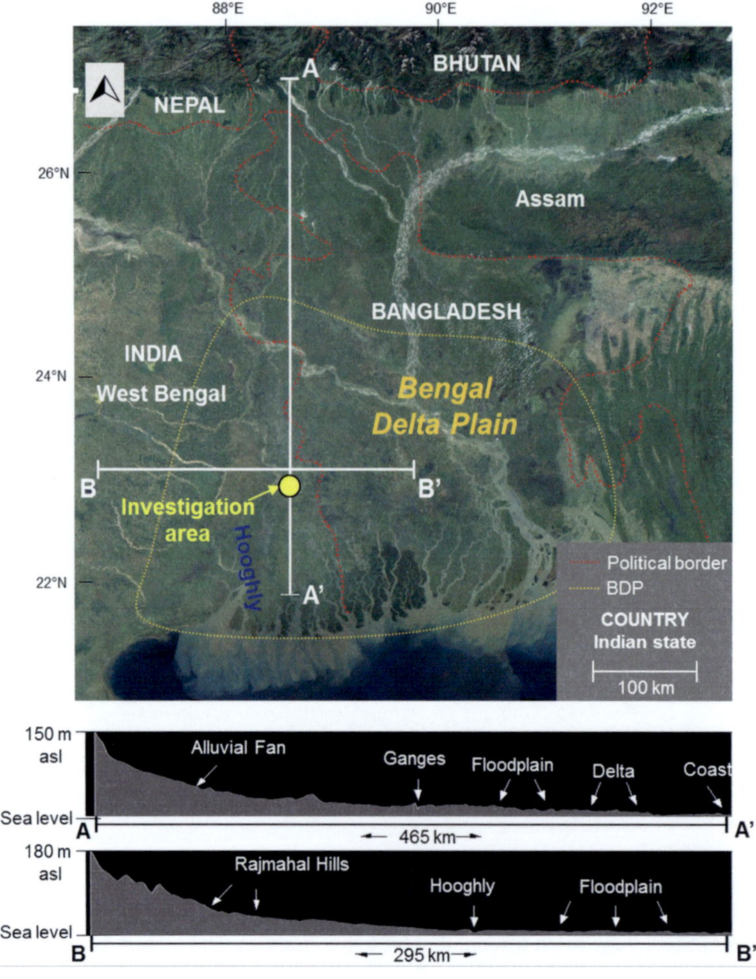

Figure 3.2: Satellite imaginary (Landsat 7) and transect elevation profiles (super-elevated) of the Bengal Basin. Image created with NASA World Wind. Resolution of satellite image is 15 m. Elevation profiles generated with NASA World Wind Terrain Profiler, based on the SRTM-30 PLUS digital elevation model (resolution: 3 arc-seconds).

3.2   GEOLOGY

The geological history of the Bengal Basin is relatively complex according to a large variety of geological processes involved in the basin formation (ALAM et al. 2003). Evolution of the proto-basin began with the break-up of Gondwanaland during the mid to late Mesozoic, followed by massive basalt extrusions at the Rajmahal and Shillong areas during the late Jurassic to the early Cretaceous (LINDSAY et al. 1991). Slow subsidence of the Bengal Shelf caused marine transgressions into the south-eastern part of the proto-basin and repeated sequences of submergence and transgression caused deposition of the first deltaic sediments. Increased tectonic activity generated a massive transgression in the middle Eocene, which covered the whole basin and formed the Sylhet limestone formation (see Table 3.1). Due to varying tectonic movement rates, evolution of the eastern and western part differed. A permanent change in the local sedimentation environment arose from the collision of the Indian and Eurasian Plates that induced uplift of the basin's stable shelf and finally formation of the Himalayan Mountains. While the eastern part remained under marine influence, the uplift changed the sedimentation environment in the rest of the basin during the middle to late Eocene, super-imposing preliminary marine carbonatic-clastic sediments by fluvio-clastic sediments (LINDSAY et al. 1991).

Modern deltaic basin evolution was initiated by intensive tectonic activities along the Dauki fault zone (ranging from east to west in front of the Shillong Plateau) and the Naga thrust zone (parallel to the Indo-Burmese Fold Belt, see Figure 3.1) during the middle and late Miocene (LINDSAY et al. 1991). A basin-wide regression followed and changed the previously marine-estuarine environment in the eastern part into a fluvial-tidal dominated system.

Table 3.1: Geology of the western part of the Bengal Basin where West Bengal is located (after ALAM et al. 2003, MUKHERJEE et al. 2009, SARKA et al. 2009).

Age	Stable Shelf, Western Basin		
	Group	Formation (thickness)	Lithology
Holocene		Alluvium (~30-60 m)	Silt, clay, sand, gravel
Pleistocene	Barind	Barind (~200 m)	Brown clay, silty clay, silty sand
Pliocene			
Miocene	Bhagirathi	Debagram / Ranaghat (~1200 m)	Sandstone, siltstone, shale
Oligocene		Pandua / Matla (~400 m)	Sandstone, siltstone, silty mudstone
		Memari / Burdwan (~150 m)	Sandy mudstone
Eocene	Jainitia	Kopili (>100 m)	Shale, mudstone, sandstone
		Sylhet Limestone (~300 m)	Fossiliferous limestone, interbedded sandstone
		Upper Jalangi (~230 m)	Sandstone, sandy mudstone
Paleocene		Lower Jalangi (~400 m)	Sandstone
	Rajmahal	Ghatal (~150 m)	Sandstone, limestone, shale
Cretaceous		Bolpur (100 m)	Sandstone, mudstone
Jurrassic		Rajmahal (~250 m)	Basalt, andesite, shale
Pre-Jurassic	Gondwana		

Beginning in the late Quaternary, basin development was mainly controlled by tectonic activities. This differs from traditional models of delta formation, which base on sea level fluctuations as controlling factor (GOODBRED et al. 2003). Tectonic movement mainly influenced the north-west of the basin close to the stable shelf, causing alternating fluvio-dynamic processes. In contrast, sedimentation at the coast was primarily affected by changes in the eustatic sea level during the ice ages, where mangrove forests and swamps followed the active coastline. Last chapter in the Modern-Delta evolution began at the onset of the last glacial maximum

(LGM) at around 20 ka BP, when the sea level was approximately 120 m lower than today. To this time, the lowest units of the delta sediments were exposed to erosion and weathering. Channel erosion incised broad valleys, while soils of several metres developed on top of the interfluves (ACHARYYA et al. 2000, ISLAM & TOOLEY 1999, LAMBECK et al. 2002).

A period of highest delta progradation followed between 7 to 9 ka BP, when changes in the regional climate intensified the seasonal monsoon precipitation. Hence, fluviatile transport and sedimentation rate of the three rivers Ganges, Brahmaputra and Meghna reached a maximum of approximately 2.4×10^9 t a^{-1} (GOODBRED et al. 2003). Due to sea-level rise at the early Holocene, the shoreline shifted again gradually in northern direction, where a huge estuary formed. Mangrove forests followed the coastline and organic-rich sediment deposit developed. During this time, fine sediments were deposited up to the northern part of the basin that filled the incised Pleistocene channels and buried the older interfluves (MC ARTHUR et al. 2008, WRIGHT & MARRIOTT 1993). After a rapid decrease, the annual sediment discharge stagnated at around 5 ka BP, but a constant sedimentation flux of about 1.0×10^9 t a^{-1} allows the active delta to gradually prograde into the Bay of Bengal to the present day (GOODBRED et al. 2003). As a result, a more fluviatile influenced floodplain environment developed in the hinterland, where thick sediment packages deposited (ALAM et al. 2003). Today, Holocene sediments can be divided into a clay- and silt-rich top, and an underlying sandy part with locally increased OM contents, especially in the southern dellta and the central floodplain (located in the western part of the Bengal Basin).

3.3   CLIMATE

The tropical monsoon climate of the Bengal Basin is mainly controlled by the southeast monsoon regime that transports moist air masses inland from the Bay of Bengal from early June to mid of October (BGS & DPHE 2001). Average air temperatures range from 10°C during winter to 35°C in summer, and annual rainfall varies between <1,200 mm in the flat western and central basin and >1,600 mm towards the Himalaya Foreland (MUKHERJEE et al. 2007). A total of 82.2 % of the annual average rainfall (AAR) falls during the monsoon season, while the dry season can be subdivided into the pre-monsoon season (between January and May with 16.2 % of the AAR) and the post-monsoon season (November to December with 1.57 % of the AAR) (MUKHERJEE et al. 2007). Extensive irrigation is supposed to exceed the absolute recharge from precipitation in parts that are under intensive agricultural use (MUKHERJEE et al. 2007). When heavy monsoon rains meet discharge water from the Himalayan snowmelt, heavy flood events occur in the flat areas of Bangladesh.

3.4   SEDIMENT FLUX AND HYDROLOGY

The annual sediment flux of the three rivers Ganges, Brahmaputra and Meghna and their tributaries was estimated as 1.0×10^9 t a^{-1}, with sedimentation rates in the BDP of about 0.5 to 1.2 m ka^{-1} since the late Holocene (SARKA et al. 2009). The mineral composition of the sediments is dominated by detrital quartz, feldspars, and minor amounts of carbonates and clay minerals (primary illite and kaolinite) (ALLISON et al. 2003, MUKHERJEE et al. 2009). Additionally, suspended matter of River Ganges generally contains high contents of smectite and carbonates, since only marginal chemical alteration appears during erosion, transportation and sedimentation (HEROY et al. 2003). River Ganges drains the Himalayan mountains and foothills and enters the basin from northwest, where he divides into two distributaries (Figure 3.1). River Padma is the main stream, which flows in southeast direction towards the confluence with the Brahmaputra in Bangladesh.

The minor part flows directly in southern direction through West Bengal, where he is called River Bhagirathi-Hooghly or simply Hooghly River. In the flat basin, the course of the Ganges is strongly meandering and has been continuously shifting eastwards for the last 250 years (MUKHERJEE et al. 2009). The Brahmaputra enters the basin in the northeast after draining Tibet and the Indian state of Assam. The Meghna River drains the Sylhet Basin and parts of the Tripura Hills, before he merges with the Brahmaputra. The active part of the delta is represented by the Sunderbans, which form a huge mangrove forest south of Calcutta (Figure 3.1).

Groundwater in the Bengal Basin is generally of the Ca-HCO_3^- -type and is mainly influenced by reducing conditions, as demonstrated by high concentrations of Fe(II) and methane (HARVEY et al. 2005). Slow ground-water flow processes enable intense water-rock interactions like carbonate dissolution, metal oxide reduction, silicate weathering as well as intrusion of saline water in areas of the southern delta. These processes have the potential to create heterogeneous and hydrochemically stratified ground-water bodies (MICHEAL & VOSS 2009a). A comprehensive groundwater survey programme conducted by the United Nations Development Programme (1978-1982) distinguished three different aquifers in Bangladesh based on a lithological stratification scheme (UNDP 1982). The uppermost water-bearing zone is represented by a shallow aquifer, situated in fine sand units below a thin confining clayey and silty unit. Below, the main aquifer is encountered in medium- and coarse-grained sandy sediments in varying depths between 75 to 140 m. In depths of 140 to 460 m, a deep aquifer represents the deepest water-bearing unit. All three aquifers are characterised by a high variability of the sediment thicknesses and locally appearing interbedded units of gravel, clay and peat.

A different approach to classify the aquifers in Bangladesh regarding their ^{14}C groundwater ages (AGGARWAL et al. 2000). In the shallowest aquifer (up to 100 m depth), groundwater age was determined as younger than 100 a. In the main aquifer between 200-300 m depth, groundwater age reaches up to 3 ka, while water residing in the deepest aquifer below is about 20 ka old.

In the Indian part of the BDP, available hydrogeological data is strongly limited, and assumptions on the local aquifers depend mainly on interpretations of lithological data and hydrochemical surveys. In the central floodplain along the Hooghly River, where the investigation area of this study is located, the aquifer has varying thicknesses of about 60 to 220 m, and comprises numerous thin clayey layers (MUKHERJEE et al. 2011). According to ^{3}H/^{3}He isotope dating, groundwater in depths of 20 m is younger than 30 a (STUTE et al. 2007).

MICHEAL & VOSS (2009a,b) raise the question if there really are multiple, disconnected aquifers within the entire Bengal Basin. Due to the heterogeneous nature of the local stratigraphy, which includes locally restricted embedded aquitards, there is no evidence for a consistent separation of distinctive aquifers at a regional scale. Despite this, basin sediments are considered to behave as one huge, hydraulically interconnected, vertically layered and anisotropic aquifer system. The occurrence of discontinuous and thin layers of fine grained sediments (clay and silt) is expected to generate anisotropic hydraulic conductivities. Aquifer pumping tests revealed that lateral conductivities (K_h) range in orders of 10^{-5} to 10^{-3} m sec^{-1}, whereas vertical conductivities (K_v) can be orders of magnitude lower (MICHEAL & VOSS 2009a,b). Based on driller log analysis, hydraulic conductivities were calculated for different regions of the Bengal Basin. In the central floodplain, the K_h was calculated as 2.2 x 10^{-3} m sec^{-1} and the K_v as 3.2 x 10^{-8} m sec^{-1} (MICHAEL & VOSS 2009b). Such a large-scale vertical anisotropy requires aquitard layers with extensions of several square kilometres and explains why local aquifers often appear to be semi-confined and disconnected.

The principal regional groundwater flow direction is from north to south, whereas the extreme flat topography of the central floodplain creates extreme slow flow rates of several decimetres per year (MC ARTHUR et al. 2008). Hence, a high vulnerability of the aquifer system arises towards perturbations caused by pumping (MICHAEL & VOSS 2009a,b).

3.5 ARSENIC IN GROUNDWATER OF THE BENGAL DELTA PLAIN

The BDP is known as one of the worst affected areas worldwide by the occurrence of partly notorious concentrations of inorganic As in local groundwater (CHARLET & POLYA 2006, POLYA & CHARLET 2009). Many studies have been conducted since the 1990's to reveal the dimension and distribution of As in local groundwater, to identify the reason of As enrichment and to provide potential mitigation strategies (e.g., BERG et al. 2001, FENDORF et al. 2010, HARVEY et al. 2002 & 2005, MC ARTHUR et al. 2001 & 2004, NEUMANN et al. 2009, POSTMA et al. 2010, VAN GEEN et al. 2004). A large field survey conducted by the British Geological Survey (BGS) and the Bangladesh Department of Public Health Engineering (DPHE) included sampling of 3,534 tube wells in the Bangladesh part of the BDP. Results revealed that 27 % of the tested wells exceeded 50 µg As L^{-1}, whereas most concerned wells are filtered in depths down to 150 m (KINNIBURGH & SMEDLEY 2001). Concentrations of total As reached up to 3,200 µg L^{-1}, with As(III) as dominating species. Groundwater with increased As concentrations was found to be characterised by high concentrations of Fe(II), HCO_3^-, PO_4^{3-} and Mn, and concomitant low contents (<1 mg L^{-1}) of O_2, SO_4^{2-} and NO_3^-. This distribution of TEA clearly points at pronounced influences of microbial Mn(IV), Fe(III) and SO_4^{2-} reduction in the shallow aquifer parts.

Elevated As concentrations typically occur in organic-enriched, grey reduced Holocene sand aquifer parts of the Ganges, Brahmaputra and Meghna delta and floodplain areas, as well as in the Sylhet Basin (ACHARYYA et al. 2000). Despite this, Pleistocene and older units are generally free of As, as well as the alluvial tract of the Ganges upstream the Rajmahal Hills (MC ARTHUR et al. 2004, MUKHERJEE et al. 2011). Typically punctual As sources like volcanic activities, thermal springs, heavy industry or mining activities could be excluded as potential sources of As (ACHARYYA et al. 2000). Arsenic concentrations can change within a couple of metres by up to three orders of magnitude (MUKHERJEE et al. 2008). Terms like "hot spots & cold spots" (CHARLET et al. 2007, NATH et al. 2008) or "patchiness" (RAVENSCROFT et al. 2001) were used to describe these inhomogeneous distribution patterns. Despite this extremely horizontal heterogeneity, concentrations of As in local groundwater also varied vertically within metres to centimetres (CHARLET et al. 2007, VAN GEEN et al. 2006). Depth profiles of dissolved As were described as bell-shaped, with maxima usually occurring in depths between 20 to 30 m below land surface (bls) (HARVEY et al. 2002). Similar inhomogeneous distribution patterns occur in other deltaic regions of Asia, too, for example in Vietnam, Taiwan or in several regions of China (CHEN et al. 1994, GUO et al. 2008, VAN GEEN et al. 2008). This high spatial variability of As is linked to the complex sedimentation history of the BDP. On the one hand, palaeo-channels and marine deposit witness an influence of regression and transgression events in dependence on the distance to the coast (SARKA et al. 2009). On the other hand, frequent river bed changes created wetlands, levees and oxbow lakes, where sedimentation caused a rapid burial of OM (sparsely distributed or concentrated in form of peat layers) (CHARLET et al. 2007, MC ARTHUR 2004 & 2008, MUKHERJEE 2009). As a result, local lithology and related geochemical and geomicrobiological processes that control the mobility of As may vary vertically as well as horizontally on a small scale.

Not much is known about the primary sources and transport mechanisms of As from the delivery areas of the Bengal Basin. Holocene sediment deposit originate from the Himalayan Mountain Range, where metamorphic rocks and magmatic intrusions form a complex geologic setting and include arsenic-bearing phases like apatite, pyrite and magnetite (HATTORI et al. 2005, MAILLOUX et al. 2009). Arsenic can be directly transported by the drainage system in fragments of As(V)- and/or As(III)-hosting minerals. In addition, As can first be mobilised in the Himalaya by weathering of surface exposed arsenic-hosting parent rock, before it adsorbs to and/or co-precipitates with Fe-(oxyhydr)oxides. These secondary Fe-mineral phases either form as thin coatings around mineral grains or as nano-particles, and act as primary carrier phases for As (RAISWELL 2011). In both cases, As(III) and As(V) are transported via surface runoff into the Bengal Basin, where they are deposited (KINNIBURGH & SMEDLEY 2001).

3.6 THE INVESTIGATION AREA AT A LOCAL TO REGIONAL SCALE

Local sites, which are located close to the study area described in the scope of this work, have been previously investigated by international research groups. During the past years, the knowledge of these sites continuously evolved based on the application of new methods and techniques. Principal results are summarised in the following and form the base to compare and incorporate new findings from this study.

3.6.1 CHAKDAH CITY ARSENIC HOTSPOTS

One of the first and currently best investigated study area in the BDP is located next to the city of Chakdah, the capital of the Nadia district. The town with its approximately 75,000 inhabitants is seated at an elevated river bank of the Hooghly River in the active floodplain of the Bengal Basin, about 60 km south of Kolkata (CHARLET et al. 2003). The geomorphologic highly dynamic area is marked by past channel shifts of the Hooghly River, which created series of oxbow lakes, meander scars, swamps, natural levees and ponds. Ponds were often used to obtain clayey building material for the traditional houses and often follow dry fallen meanders.

From 1998 to 1999, 636 deep, 245 medium deep, and 587 shallow tube wells have been used together with 319 river lift pumps to irrigate approximately 112,089 hectares of land in the Nadia district, which is about 78 % of the locally cultivated land (NATH et al. 2008). Since then, irrigation based on shallow groundwater has furthermore increased.

Groundwater from local tube wells and newly installed monitoring wells reflected a characteristic groundwater composition, including pronounced spatial variations of As concentrations that reached up to 500 µg L^{-1} and appeared in form of plumes (CHARLET et al. 2003 & 2007). High ground-water HCO_3^- contents (compared to Hooghly River water) in the shallow groundwater were attributed to the microbial degradation of considerable amounts of OM (MÉTRAL et al. 2008). According to E_H measurements and hydrochemical compositions of groundwater with increased As contents, Fe(III) reduction was identified as one of the predominating TEA consuming processes. In order to determine the spatial extent of a typical As hot spot, transects reaching from the Hooghly River banks towards the city were sampled with the needle sampler technique developed by VAN GEEN et al. (2006). Interpolated results reflected a partly decoupling of groundwater As concentrations from dissolved Fe, and a distribution pattern mirror-inverted to that of SO_4^{2-} (MÉTRAL et al. 2008). Two As plumes were described with spatial extensions of only a few hundred metres and thicknesses of about 10 to 30 m. The spatial extension of a several metres thick surface clay layer that overlies the sandy aquifer could be determined by combining drillings and electromagnetic conductivity mappings. This hydraulically nearly impermeable layer turned out to be patchy, comprising pockets of sandy sediments. In groundwater below these sandy pockets, As, Fe, and major ion concentrations turned out to be low, while SO_4^{2-} was increased (MÉTRAL et al. 2008). These patterns were attributed to locally restricted recharge with vertically infiltrating, oxygen- and nitrogen-rich surface water, which keeps the system constantly in a state of less reducing conditions.

Piezometric measurements reflected seasonal, vertical fluctuations in the hydrostatic head of about five metres, whereas the Hooghly River acts as recharge during the dry season, and as drainage during the monsoon (NATH et al. 2008). Deep and shallow aquifer zones are intra-connected as demonstrated by influences of a newly installed governmental deep drinking water well in the eastern part of Chakdah. Excessive withdrawal of groundwater in depths of 100 m and deeper created a local depression cone in the surface near shallow aquifer, causing subsequent attraction of a nearby As plume and drawdown of arsenic-rich groundwater in range of the deep supply well (CHARLET et al. 2007, LAWSON et al. 2008).

Geochemical characterisations confirmed the outstanding role of Fe-(oxyhydr)oxides as primary carrier phase of the predominantly low As contents (around 2.5 mg kg^{-1}) and the prevalence of As(III) (>70 %) in the reduced sediments (ROWLAND et al. 2005). Geo-microbiological cultivation experiments with sediment samples from a field site near the Hooghly River provided important insights into the biogeochemical cycling of As.

Indigenous microbial communities were found to be very diverse, covering nine classes, six phyla and many unknown species (HÉRY et al. 2008, HÉRY et al. 2010).

The following species could be identified demonstrating the close relation between Fe(III) reduction and As release:

- *Sulfurospirillum spp.*, capable of Fe(III) and As(V) reduction (ROWLAND et al. 2009);

- *Acinetobacter spp.*, As tolerant and known to release As from wood preservatives (GAULT et al. 2005);

- *Geobacter spp.*, generally capable of Fe(III) reduction (ISLAM et al. 2004), with the *arrA*-gene detected in *Geobacter uraniireducens* and *Geobacter lovleyi* (HÉRY et al. 2010).

Laboratory experiments with isolated strains of these indigenous strains successfully demonstrated the ability of Fe(III) reduction and release of dissolved As(III) and Fe(II) (ISLAM et al. 2004). Further experiments revealed a partial decoupling of As mobilisation from Fe(II) release. This was attributed to As re-adsorption following transformations of Fe(III)-mineral phases by dissolved Fe(II) and/or the presence of DARPs (GAULT et al. 2005, ROWLAND et al. 2009). Although peat layers are supposed to be closely linked to the occurrence of arsenic-enriched redox zones, no peat was found in local aquifer sediments (CHARLET et al. 2007, ROWLAND et al. 2006). Sedimentary organic matter was characterised as strongly biodegraded, not of anthropogenic origin, increased in high molecular weight hydrocarbons (n-alkanes) and still able to support microbially mediated As(III) release in microcosm experiments (ROWLAND et al. 2006). Results alluded to the conclusion that release and distribution of As in groundwater of this area is driven by microbial activity in anoxic aquifer parts, which is in turn influenced by local geomorphologic characteristics and pumping (CHARLET et al. 2007).

3.6.2 *INFLUENCES OF THE LOCAL SEDIMENT STRATIGRAPHY*

Another long and well investigated study site is located at the junction of three neighbouring villages Joypur, Ardevok and Moynar (abbreviated as JAM), about 38 km south of Chakdah and 12 km east of the Hooghly River. Shallow groundwater compositions met in local tube wells turned out to be typically for the Bengal Basin, with As concentrations reaching up to 1,180 µg L^{-1} (MC ARTHUR et al. 2004). Field survey results revealed that local groundwater is compositionally stratified by depth, with strong variations of redox sensitive parameters like As, Mn, Fe, and SO_4^{2-} that formed so called "redox fronts" (MC ARTHUR et al. 2008). These redox fronts reflect a downward decreasing redox potential in the local aquifer. Faecal OM could be excluded as potential source for organic carbon triggering As release (MC ARTHUR et al. 2004). Age determination of the local aquifer sediments revealed that in 31.5 m depth Holocene (~7,000 a BP), and in 50 m depth Pleistocene (~27,230 a BP) sediments occur (MC ARTHUR et al. 2004). Despite this, respective groundwater turned out to be very young with ages ranging from 1.4 to 32.2 a in the Holocene aquifer and >50 a in the Pleistocene parts (MC ARTHUR et al. 2010). Lateral groundwater flow rate in the Holocene shallow aquifer part was calculated from piezometric data as ~0.1 m d^{-1} (~30 m a^{-1}), while downward velocities ranged in dependence of local irrigational pumping between 0.05 and 0.60 m d^{-1} (MC ARTHUR et al. 2008).

Some of the nested monitoring wells were installed in sandy palaeo-channel sediments, while others were suited in sediments of a palaeo-interfluve, including buried remains of an oxidised palaeosol. The palaeosol covers an area of about 400 x 450 x 5 m, is composed of brown clay holding up to 299 mg As kg^{-1}, and is underlain by Pleistocene brown sand (MC ARTHUR et al. 2010). It was proposed that this soil had developed during the last sea-level lowstand in time of the last Pleistocene glacial maximum (LGM), before it was subsequently buried by Holocene deposits.

Buried palaeosols are considered as a widespread geomorphologic feature in the BDP, whose Fe(III)-rich brown sands function as effective As barriers and protect underlying aquifers from sinking arsenic-rich groundwater (MC ARTHUR et al. 2011, STOLLENWERK et al.2007).

Annual sampling in the monitoring wells revealed diverse temporal changes in As concentrations between 2001 and 2009, varying from decreasing over peaking to increasing trends (MC ARTHUR et al. 2010). These trends were attributed to (i) flushing and infiltration of arsenic-low groundwater; ii) downward movement of the chemically-stratified water column; (iii) and lateral invasion of arsenic-enriched water in zones of brown sand (MC ARTHUR et al. 2010). Excessive local abstraction of groundwater for irrigation and drinking water supply has strongly influenced the local groundwater flow directions and caused temporal trends in dissolved As concentrations of some monitoring wells, although the occurrence of high As groundwater generally predates pumping (MC ARTHUR et al. 2010).

3.6.3 CONCEPTIONAL MODEL OF ARSENIC RELEASE ACCORDING TO PRESENT KNOWLEDGE

Arsenic release within groundwater of the BDP is an active process, which is controlled by microbial redox reactions as well as related abiotic processes. Resulting As distribution patterns in groundwater are influenced by regional and local hydrogeological conditions and artificial groundwater abstraction. Arsenic itself can be considered as a passive element, which does not directly influence the controlling factors. The complex interactions involved in biogeochemical cycling of As in the BDP can be summarised in a conceptual model (Figure 3.3).

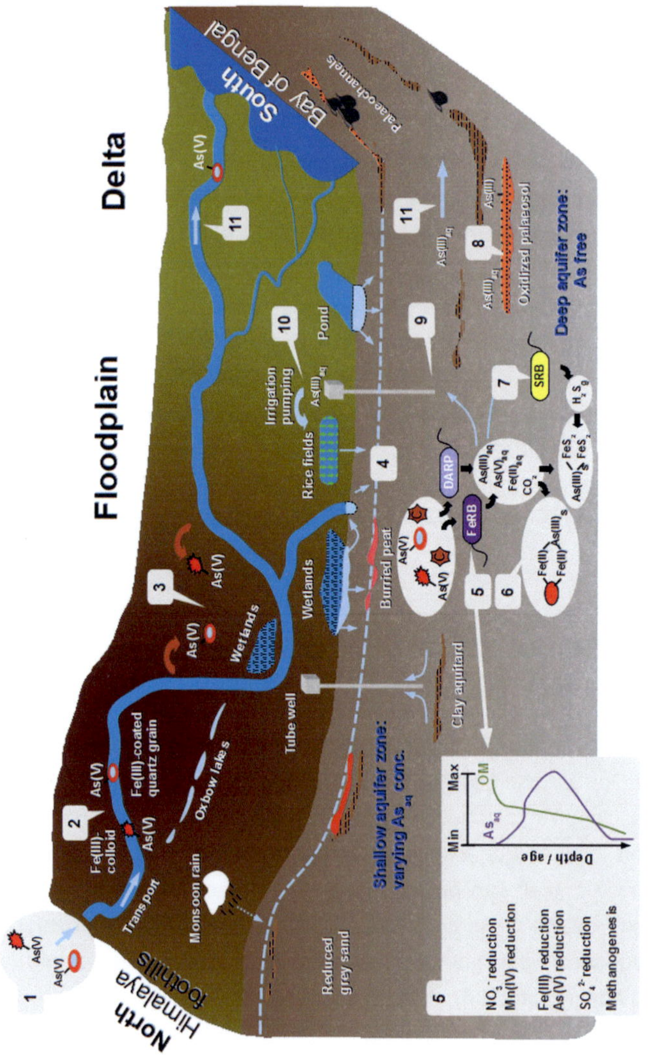

Figure 3.3: Conceptual model of the As cycle in the Bengal Basin after KINNIBURGH & SMEDLEY (2001). Figure caption continued on next page.

Figure 3.3, continued:

1. Release of As in the Himalaya region by weathering of parent rock followed by immobilisation through secondary Fe-(oxyhydr)oxides;

2. Fluvial transport of fragments of arsenic-bearing primary minerals and secondary arsenic-hosting Fe-(oxyhydr)oxides;

3. Continuous burial of arsenic-bearing minerals together with high amounts of organic matter;

4. Surface recharge from ponds, wetlands and irrigated fields provides infiltration of fresh water (enriched in DOC and TEA) through sandy gaps in the surface aquitard;

5. Development of vertically and horizontally stratified redox zones in the aquifer sediments depending on the availability of OM and TEA. Interconnected biotic and abiotic processes affect the mobility of As, when the redox potential successively declines in shallow aquifer parts;

6. + 7. Arsenic re-adsorption onto transformed Fe-minerals (6) and precipitation of arsenic-retaining Fe-sulphides (7) result in extremely varying groundwater As concentrations;

8. Slow and anisotropic transport of dissolved As and adsorption onto Fe(III)-oxides of buried palaeosols, which protect the underlying aquifer parts;

9. Excessive groundwater abstraction redirects natural flow paths, which induces mixing of water with different hydrochemical properties and endangers arsenic-free groundwater;

10. Irrigation water causes increase in soil As contents and partially re-import of As into the aquifer via infiltration;

11. Flush-out of As via surface and groundwater discharge in southern direction into the Bay of Bengal.

It is not the question if there is As release in the BDP, but how, where and when it occurs. Though considerable progress has been made, a huge number of questions remains unsolved that will be addressed in the following chapters:

- The ultimate cause for the development of small-scaled redox zonings in the aquifer sediments remains obscure and there is in addition a lack of necessary data to discern the relative importance of other than microbial mechanisms related to As release and enrichment (VAN GEEN 2011);

- Regarding the widely accepted Fe(III) reduction hypothesis, it remains unclear why no steady re-adsorption of As onto surfaces of residual Fe-minerals appears (WELCH et al. 2000). This critical question appears to be a concealed aspect in present literature reporting about As enrichment in the BDP as well as other affected Asian deltas and floodplains;

- Further, little is known so far regarding the recently discovered transformations of Fe-minerals, which influences the sedimentary As retention potential (KOCAR et al. 2006, PEDERSEN et al. 2006);

- In present literature, nearly no attention is paid to processes in the groundwater fluctuation zone close to the surface, where redox transformations at the solid-water-interface appear in scales of µm to cm;

- Importantly, it is of greatest relevance to answer how vulnerable shallow arsenic-free aquifers are really to pumping, since they are the currently preferred source of drinking and irrigation water in the BDP (MICHAEL & VOSS 2009a, NATH et al. 2008).

4. MATERIALS AND METHODS

4.1 MULTILEVEL WELL INSTALLATION AND SEDIMENT SAMPLING

Five nested multilevel wells were installed at two study sites entitled as the high As site and the low As site during June 2008. Each well was installed in a separate borehole, with a maximum distance of ~3 m to a central well in the middle. Well screens and casings made of PVC were placed in different depth ranges (well screen positions low As site: 12-21, 24-27, 30-33, 36-39 and 42-45 m bls; high As site: 12-21, 22-25, 26-29, 30-33, 34-37 m bls). Gravel surrounds the well screens (pore size of the screens: 2 mm) as filter pack and overlaying, compacted bentonite serves as sealing (see Figure 4.1). The lockable tube wells are additionally sealed with concrete at the surface to prevent infiltration of surface water. Below each well screen, a mud trap of 1 m length is located. The central well A of each site has a diameter of 20 cm, while the surrounding wells are equipped with 7.5 cm well screens and casings. At both sites, installation of the deepest wells was combined with sediment sampling. A split-spoon core barrel lined with a PVC tube (length: 60 cm, diameter: 5 cm) was used as core catcher (KIEFT et al. et al. 2007). The core barrel was locked in front with a cutting shoe, providing additional 5 cm of core samples that were retained separately. The system was attached to a drilling rig and hammered manually into the unconsolidated aquifer sediments with a 50 kg weight (ISLAM et al. 2004). After sampling of 1.30 m (2 x 0.65 m), drilling for the well instalment was made up for the respective depth interval, before the next interval was sampled and so on.

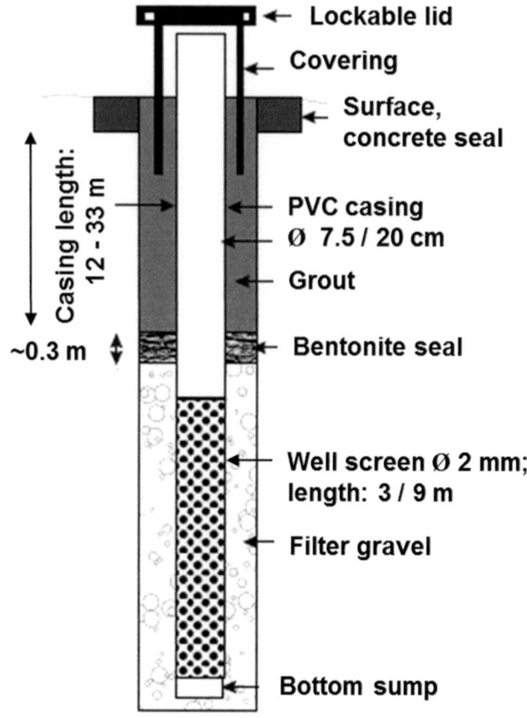

Figure 4.1: Construction details of the monitoring wells according to LAPHAM et al. (1997).

The challenge was to obtain uncompromised, representative sediment samples for microbiological investigations and column experiments. Thus, the use of drilling fluid was avoided as long as possible and the equipment was cleaned with a disinfecting agent (Sterillium®, BODE CHEMIE) before sampling. First, hollow stem augering was used until the saturated zone was reached. Then, drilling was continued with a cable tool drilling rig until a depth of about 11 m bls (low As site), respectively 13 m bls (high As site). Finally, drilling was completed with rotary drilling, using bentonite solution as drilling fluid. Captured sediment cores (average recovery: 64.4 %) were immediately purged with N_2 and sealed with a PVC cap and air tight tape to minimize contact with O_2. Uncompromised cores with aquifer sediments and pore water were separated and used for microbial column experiments. Sample material from the cutting shoe was rapidly transferred into nitrogen purged HDPE bags, sealed twice and stored under N_2 atmosphere in two PVC barrels to avoid oxidation. The tightly closed barrels were regularly (once per week) purged with N_2 until sample analysis.

After installation of the monitoring wells, well development was done according to common practice (LAPHAM et al. 1997). Unfortunately, the central wells (12-21 m bls) collapsed at both sites and were rebuilt in December 2008. The two new wells were finalised in December, shortly before the first sampling.

4.2   GEOCHEMICAL SEDIMENT CHARACTERISATION

Sediment analyses aimed to gather information about the aquifer mineralogy, geochemistry and OM in general, and the arsenic-bearing mineral phases in particular. All measurements were carried out at the IMG, except for TN determinations (done at the Institute of Geography and Geoecology, IfGG). Sample treatment procedures are summarised in Figure 4.2. Samples were measured in replicates and blank measurements were included for blank value correction and to calculate respective limits of detection (ld). The same applies to hydrochemical analyses (chapter 4.3).

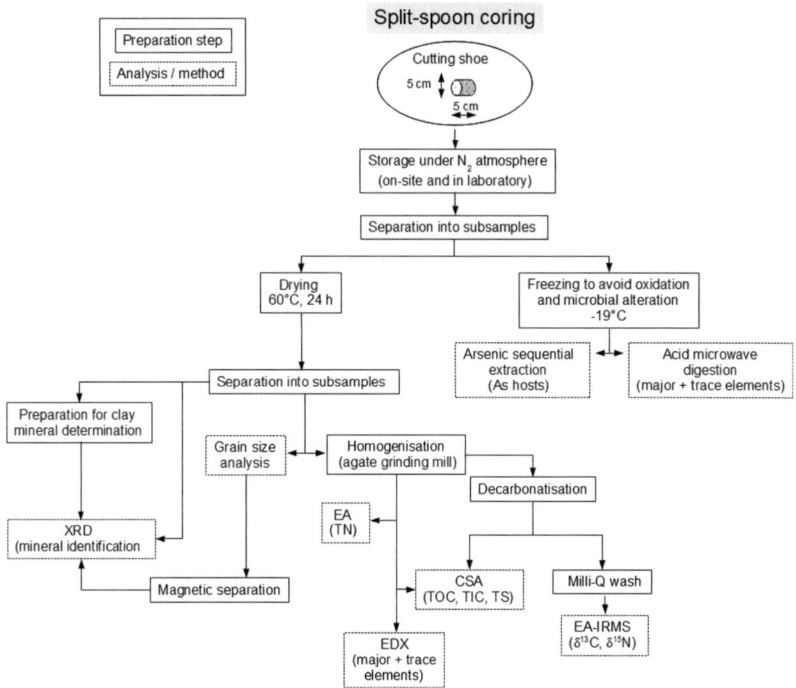

Figure 4.2: Flow chart representing sample preparation and sediment analyses. Abbreviations given in text below.

Sediment analysis. Grain size analysis was done following DIN 4188, differentiating down to the silt and clay fraction (grain size <63 µm). Clayey samples were wet-sieved, sandy samples dry. Results were used to estimate hydraulic conductivities (K) of respective aquifer parts with the Beyer formula (BEYER 1964). Samples were dried (60°C), homogenised in an agate grinding mill and analysed for major and trace elements by energy dispersive X-ray spectroscopy (EDX; Epsilon 5, PANanalytical). Results were corrected to average deviations from certified values of reference materials (see APPENDIX I, Table A 1.1). Accuracy and precision were generally very good.

For example, relative standard deviations (%RSD) for As in the reference material GXR-5 was 2.78 % (average content: 12.2 mg kg^{-1} As, n: 10), and in case of a low As sample 1.84 % (average content: 2.16 mg kg^{-1}, n: 3).

Representative samples from different depths were chosen for detailed analyses. Microwave assisted acid digestions were conducted (following MLS application E704 for rock and sediment) to determine additional elements (Na, Mg, P and V) and to validate EDX and sequential extraction results (see below). For each measurement, previously frozen sediment material (about 50 mg) was transferred into microwave Teflon flasks (MW system Start 1500, MLS GmbH) and the exact weight noted. Then, 0.5 mL H_2O_2 (30 % p.a.; ROTH), 0.25 mL HF (40 % suprapure, Merck), 2.0 mL HNO_3 (65 % sub-boiled, Fluka) and 1.0 mL of Milli-Q water were added. After digestion, the solution was transferred into small Teflon cups and evaporated at about 120°C to the size of a small drop. Residues were diluted with HNO_3 (1 % sub-boiled, Fluka) and evaporated again. After two repetitions, digestions were transferred into 10 mL volumetric flasks and stored in the refrigerator at 4°C until elemental analysis. This was done in a cleanroom by means of inductively coupled plasma mass spectrometry (ICP-MS; XSERIES 2 ICP-MS, Thermo Fischer). The GXR-2 reference material was included as reference material, see section "sequential extraction" below.

Mineral phases were qualitatively identified by powder X-ray diffraction (XRD; D500 Kristalloflex, Siemens; measuring quality and reproducibility of reference materials and samples summarised in APPENDIX I, Table A 1.2). Detection limits for accessory minerals were improved by stepwise separation and pre-concentration of para- and diamagnetic minerals using a magnetic separator (Isodynamic separator model 2, Frantz; settings summarised in APPENDIX I, Table A 1.3). Sample material (30 g) was dry-sieved to separate the fraction between <2 mm and >0.063 mm, which was weighted and further separated by magnetic separation. Before, highly magnetic magnetite was qualitatively detected with a bar magnet. Clay minerals were determined via XRD from clay-rich samples (nine samples from the high As site and six samples from the low As site) by comparing powder X-ray diffractograms (2 to 22° 2θ) of untreated samples to glycol treated (24 h) and heat exposed (550°C, 1 h) subsamples. All XRD measurements were carried out with CuKα radiation (40 kV and 25 mA) in steps of 0.01° 2θ s^{-1}. Obtained spectra were processed (background subtraction, k-alpha stripping and minor smoothing) with PowderX (beta version, developed by Cheng Dong) before data was evaluated with DIFFRAC PLUS EVA (version 13.0.0.1, Bruker AXS) and Qual-X (version 1.2, Instituto di Cristallografia Italia), which both use the PDF-2 database (version 2004, ICDD).

Sequential extraction procedure. In order to identify the host phases of sedimentary As in sediments, a sequential extraction procedure (SEP) was performed according to EICHE et al. (2008 & 2010) and WAGNER (2005). This extraction is a modification of the SEP developed by KEON et al. (2001), VAN HERREWEGHE et al. (2003) and WENZEL et al. (2001). The detailed procedure is summarised in APPENDIX I, Table A 1.4. Results of samples analyses of the study sites are presented in chapters 6 and 7 (Figures 6.4 and 7.3). In 3 out of 17 extractions, the sum of extractable As was considerably lower than total As (As_{tot}) contents determined by EDX and acid microwave digestions, while the other samples showed a moderate agreement. A certified reference material (GXR-2) was analysed, too. The measured As_{tot} content of 28.8 mg kg^{-1} is in accordance to results of microwave acid digestions (28.4 and 28.2 mg kg^{-1}), but EDX measurements (24.9 ± 0.8 mg kg^{-1}; n: 11) as well as the certified value (25.0 mg kg^{-1}) are lower. Additionally, two samples were extracted in duplicate. Although total As contents are similar (high As site: 4.73 and 4.89 mg kg^{-1}; low As site: 1.49 and 1.63 mg kg^{-1}, respectively), the relative portions in fractions I to VII appear to be shifted, especially in fraction II (strongly adsorbed As).

Increased contents of dissolved Fe occurred in fraction II, which is attributed to an early dissolution of iron phases (see APPENDIX II, Table A 2.2 and APPENDIX III, Table A 3.2). The long (16 & 24 h) and repeated leaching with 0.5 M NaH$_2$PO$_4$ solution caused most likely dissolution of amorphous Fe-(oxyhydr)oxides, although the underlying mechanism remains unclear. This results in an overestimation of As contents in SEP fraction II at the expense of fraction III, which includes very amorphous Fe-(oxyhydr)oxides. However, an increased proportion of As in the strongly adsorbed fraction is still plausible according to high contents of dissolved As in local groundwater, prolonged water-rock interaction and high affinities of Fe-(oxyhydr)oxide surfaces for As(III) and As(V) (DIXIT & HERING 2003).

The reliability of this method was proved by EICHE et al. (2008), where differences in replicates were less than 10 %. Differences in As_{tot} contents between sum of the SEP fractions, EDX and acid microwave digestions as well as deviant As distribution patterns in replicates are attributed to inhomogeneities within the untreated sample material and the generally low As contents. Furthermore, it is difficult to distinguish the role of specific oxides from extraction data due to the limited selectivity of the extractants (KINNIBURGH & SMEDLEY 2001). Thus, results of the SEP must be treated with caution and considered as semi-quantitative.

Determination of iron minerals. Iron contents (determined as Fe_2O_3) of 1.36 to 5.83 weight % (wt.%) indicate the presence of considerable amounts of iron-bearing minerals, but crystalline Fe-(oxyhydr)oxides could not be directly identified by XRD (ld: ~5 %).

Semi-quantitative estimations can be made from the different SEP fractions, but results must be considered as underestimated regarding the previously mentioned loss of Fe in fraction II:

- Fraction III (HCl): Iron bound in sulphides, carbonates and very amorphous Fe-(oxyhydr)oxides (KEON et al. 2001);

- Fraction IV (AOD): Iron in form of amorphous Fe-(oxyhydr)oxides (CORNELL & SCHWERTMANN 2003);

- Fraction V (DCB): Most well crystallised Fe(III)-oxides (PANSU & GAUTHEROU 2006).

More precise methods for direct detection of arsenic-hosting phases as micro-synchrotron (µS) based techniques (µS-XAFS and µS-XRD in combination with µS-EDX) were extensively tested at the SUL-X beamline of the ANKA facility. These methods turned out as inadequate due to the overall low As contents.

Characterisation of organic matter. Total S (TS; ld: 26 mg kg^{-1}) and C (TC; ld: 71 mg kg^{-1}) contents were determined with a carbon sulphur analyser (CSA; CS 2000 MultiLine F/SET-3, Eltra). Total organic C (TOC) and inorganic C (TIC) were distinguished by comparing untreated sediment samples with decarbonised (with 20 % HCl p.a., Merck, at 60°C) ones. Quality of the analyses was controlled by repeated determinations (per 10 samples) of two reference materials (see APPENDIX I, Table A 1.5). Total N contents were measured in triplicates with an elemental analyser (EA; EuroEA3000, Euro Vector). Results of the two regularly measured reference materials were constantly too high (see APPENDIX I, Table A 1.5), whereas results were batch-wise corrected to the certified value of the standard NCS DC 73326.

Analytical limitations arose for stable isotope measurements of δ^{13}C from the generally low TOC contents that were previously determined by CSA. Thus, isotopic analyses were restricted to a subset of 49 samples (high As site: 27; low As site: 22). Isotopic δ^{13}C (measured against VPDB) and δ^{15}N (measured against atmospheric N) values were determined with an EA (EuroEA3000; Euro Vector) coupled online to an isotope ratio mass spectrometer (IRMS; IsoPrime, GV Instruments). The δ-values were calculated according to HOEFS (2009). Isotopic values were determined from previously decarbonised TOC samples. To remove unwanted remains of Cl$^-$, samples were repeatedly washed with Milli-Q water in glass bottles before drying at 40°C. Accuracy was checked with certified reference materials in regular intervals (see APPENDIX I, Table A 1.5).

4.3 HYDROCHEMICAL MONITORING AND IN-SITU EXPERIMENTS

4.3.1 FIELD SURVEY AND REGULAR MONITORING

Field survey. The aim of the preliminary survey was to obtain a detailed picture of the local aquifer hydrochemistry in the proposed study area within the Nadia district. Between September and November 2007, 174 private and governmental wells were sampled in an area of approximately 20 km^2. The study area is located approximately 80 km north of Calcutta (Figure 5.1). Sampling of local private tube wells was conducted by two Indian PhD students (A. BISWAS and S. MAJUMDER) from the Department of Chemistry of the University of Kalyani (West Bengal, India). During sampling, field parameters (electrical conductivity, pH, water temperature) were measured on-site with a multi meter (MultiLine F/SET-3, WTW). Groundwater samples were collected after field parameters had stabilised to assure that well water was completely exchanged. Results of this survey were used to identify two capable field sites with contrasting As contents in shallow groundwater. The detailed decision-making process is discussed in chapter 5.3.2.

Regular monitoring. The monitoring wells enabled sampling and monitoring of the groundwater chemistry in five different depth ranges at each study site. Between December 2008 and August 2010, sampling was done in bi-weekly intervals by A. BISWAS and S. MAJUMDER. Samples were taken with a submersible electrical pump system (delivery rate: 2.4 L min^{-1}) after stabilisation of field parameters. The low pump delivery rate was applied to avoid mixing of water from different depth ranges.

Undisturbed conditions were recorded during the first year, while in 2010 short- and middle-term influences of the in-situ experiments were monitored. Before sampling, undisturbed hydraulic heads were recorded with a light plummet. Since no exact elevation data was available, the top edges of the monitoring well tubes were levelled against each other using a water filled hosepipe to obtain a comparative point of reference for hydrostatic head measurements. Additionally, the tube well height above the ground level was measured. Immediately after sampling, dissolved Fe(II) and total Fe concentrations were determined spectrophotometrically (Lambda 20 UV-VIS Spectrophotometer, Perkin Elmer) with the phenanthroline method (SANDELL 1959) at the Kalyani University. Samples were shipped regularly to the Karlsruhe Institute of Technology (KIT), where the hydrochemical analyses were carried out at the Institute of Mineralogy and Geochemistry (IMG). Organic and microbiological analyses were done at the Institute of Biology for Engineers and Biotechnology of Waste Water Treatment (IBA) by D. FREIKOWSKI.

4.3.2 *IN-SITU FIELD EXPERIMENTS*

After one year of monitoring, in-situ experiments were conducted between 04/12/09 to 18/12/09. In addition to measurements done during the regular monitoring, total alkalinity (TA) was determined on-site using a titrimetric test kit (alkalinity test 1.11109.0001, Merck; %RSD: ± 2.00, n: 5). Groundwater O_2 concentrations and E_H were measured with respective multi meter probes inside a glassy flow cell to avoid influences of atmospheric O_2. Unfortunately, only a few reliable E_H measurements are available because the probe broke soon. Generally, all probes were calibrated before use. The undisturbed hydrochemistry was recorded at each site one day before the experiments were conducted (03/12/09). Before and at the end of the experiments, NH_4^+ was measured in triplicates spectrophotometrically (NH_4^+ test kit for Spectroquant$^®$ 0.26-10.30 mg L^{-1}, Merck). Details of sample treatment are included in Figure 4.3.

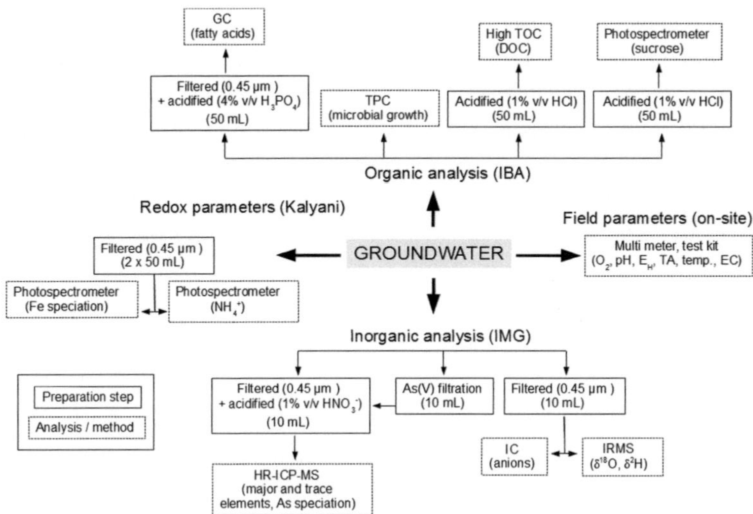

Figure 4.3: Treatment and hydrochemical analyses of samples taken during the in-situ field experiments.

The field experiments were supported by the Indian project partner regarding permissions, logistics and completion. The regular monitoring was continued during the next seven months to monitor mid-term effects.

In-situ biostimulation experiment. At the low As site, the microbial potential of releasing As by anaerobic degradation of OM was investigated. Therefore, sucrose (saccharose) was introduced into the local aquifer as an easy biodegradable and non-reducing organic carbon source (PÉREZ 1995). Sucrose was dissolved in 30 L of previously extracted groundwater and added stepwise into the four surrounding wells B, C, D and E. Four kg of dissolved sucrose was each introduced into wells B and E, and two kg into wells C and D, respectively. Importantly, sucrose was added only once and no further substances (especially TEA) were introduced. To create a diffuse distribution of sucrose within the aquifer by avoiding insertion of O_2, circular pumping was used. To do so, groundwater was drawn from the central well A with an electrical pump (delivery rate: 70 L min^{-1}) and was directly re-introduced into the surrounding wells, beginning with the deepest well E (Figure 4.4). To assure that the extracted and re-introduced groundwater contained sucrose, a simple qualitative colorimetric field test using $CoNO_3$-solution was done every minute (according to WHITEHEAD & BRADBURY 1950). After three minutes, the test indicated the presence of sucrose in delivered groundwater. Thus, sucrose was successfully introduced and locally restricted distributed in the aquifer via circular pumping. This step was repeated for wells B, C and D at the 04/12 and 05/12/09. For each well, a volume of 1,200 L was circulated. To increase the hydrostatic pressure during re-injection, a 200 L barrel was mounted on top of wells B and E. Assuming a slow groundwater flow velocity, the generated sucrose cloud was expected to last at the study site for the next couples of weeks. After infusion, samples were taken from all wells every second day (06/12 to 18/12/09).

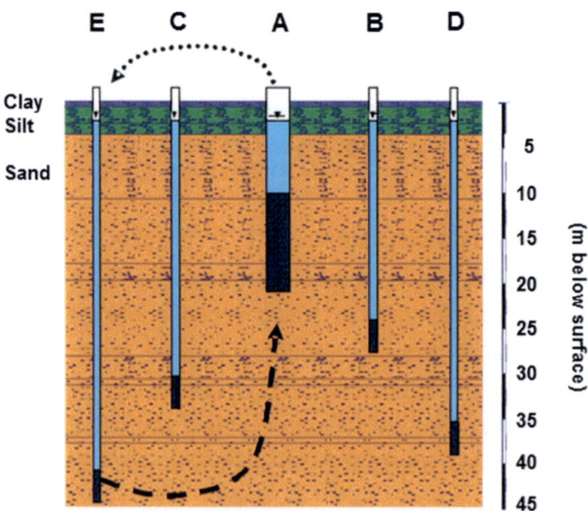

Figure 4.4: Schematic sketch of step-wise sucrose injection by circular pumping. The procedure was repeated for the four surrounding wells B to E.

58

Groundwater abstraction experiment. At the high As site, a groundwater table drawdown experiment was conducted to simulate excessive groundwater abstraction and to aerate surface near, permanently reducing aquifer sediments (setup illustrated in Figure 7.9, chapter 7.2.2.2). The pump used for the experiment (borewell pump model 100W15RA3, Crompton Greaves LTD) was purchased in Chakdah city and is a common model used by local farmers for irrigation purposes. It has a diameter of 100 mm, 2.2 kW performance and a maximum discharge of 150 L min^{-1} down to 50 m depth. The pump was installed at a depth of 15 m inside the central pumping well A (well screen covering 12-21 m bls). This is a common depth range used for local tube wells (NATH et al. 2008). After installation, four pump and rest cycles were completed between 04/12/09 and 17/12/09. Each cycle included 48 h of continuous pumping, followed by a rest cycle of 24 h. Accompanying sampling was done after each rest cycle to include potential ion exchange reactions, which may have subsequently affected the mobility of As. For comparison, one additional sampling was done immediately after the third pumping interval before the rest cycle. In total, a volume of about 1,730 m^3 was abstracted and distributed among surrounding ponds and fields. To avoid re-infiltration of discharged water, a minimum distance of 150 m to the pumping site was maintained.

4.3.3 *GROUNDWATER ANALYSIS*

Major and trace elements. Major and trace elements (including As and P) were determined from filtered (0.45 µm cellulose acetate filter, Sartorius) and acidified (1 % v/v HNO$_3$ suprapure, Merck) samples by high resolution ICP-MS (HR-ICP-MS; AXIOM, VG Elemental) in a cleanroom (APPENDIX I, Table A 1.6). The analytical precision was estimated by multiple sample replicates (triplicates), while the accuracy was assessed by analysing two certified reference materials every seven samples (APPENDIX I, Table A 1.7). In addition, samples for As speciation were additionally percolated through an As(V) retaining filter cartridge (MetalSoft Centre, MENG et al. 2001). Net differences between total As and As(V) removed samples are interpreted as inorganic As(III), whereas the presence of organic species like MMAA and DMAA are considered negligible according to previous speciation studies done in the BDP (CHATTERJEE et al. 1995, DAS et al. 1996) and in the black foot disease area of Taiwan (CHEN et al. 1994, LIN et al. 1998). Constantly high ratios of determined As(III) (>90 %) that are in good agreement with thermodynamic species calculations done with PHREEQC (see below) demonstrate that As(III) oxidation did not occur in samples during storage. Limits of detection were calculated for each sample batch separately from the standard deviation (σ) of blank values that were included every seven measurements (APPENDIX I, Table 1.6).

Anion concentrations (Cl^-, NO_3^- and SO_4^{3-}) were determined by ion chromatography (IC; ICS 1000, DIONEX) from filtered sample aliquots, except for PO_4^{3-}. Quality control was done with regular measurements of a reference solution containing comparable anion concentrations (SPEC PURE^{TM} Multi Ion IC Standard Solution, Alfa-Aesar). Limits of quantification (lq) strongly depend on the column and sample volumes used. During the monitoring period, two different separation columns were used (AS 4 SC and AS14, both DIONEX). Hence, quantification limits were determined for each column separately (according to DIONEX Application Note 133) based on regularly included control measurements (APPENDIX I, Table A 1.8).

From anion samples, δ^2H and $\delta^{18}O$ signals were obtained by Gasbench-IRMS (Delta V Advantage, Thermo SCIENTIFIC) and standardised to the VSMOW reference material (COPLEN et al. 2000). Analysis quality was regularly checked with three certified reference solutions (see APPENDIX I, Table A 1.9).

While anions measured by IC are given with respective ionic charge, major and trace elements determined by HR-ICP-MS are given in non-ionic form if not stated otherwise. Although these elements are definitely present in ionic form, it is common practice in the As research community not to state the ionic charge (e.g., CHARLET et al. 2007, MC ARTHUR et al. 2010, VAN GEEN et al. 2006). In order to distinguish between results of the speciation analysis for As and Fe, species concentrations are given as Fe(II) and As(III).

In some cases thermodynamic species calculations were done with PHREEQC (described below). To avoid confusion, these calculated species are stated additionally when used in tables and figures. The only exception is P. Phosphate was calculated from P contents measured by HR-ICP-MS, assuming that dissolved P occurs in orthophosphate (PO_4^{3-}) form (DATES 1994). This step is necessary because present groundwater is reduced. Otherwise, Fe-(oxyhydr)oxides precipitate in non-acidified anion samples and remove thereby dissolved PO_4^{3-}. For example, one acidified sample was compared to a non-acidified sample, both analysed by HR-ICP-MS. Dissolved Fe declined from 5.57 to 0.19 mg L^{-1}, concomitantly removing dissolved As (from 151 to 74.5 µg L^{-1}) and PO_4^{3-} (3.54 to 0.18 mg L^{-1}). This is an impressive demonstration how effective Fe-(oxyhydr)oxides can immobilise PO_4^{3-} and As(III) (representing 98.5 % of total As in this sample) by adsorption and incorporation.

Organic and microbial analyses. Organic and microbial analyses were done by D. FREIKOWSKI at the IBA. Dissolved organic carbon (DOC) was determined from acidified samples (1 % v/v HCl suprapure, Merck) with a TOC analyser (HighTOC, Elementar). In addition to DOC analysis, further determinations were done for samples obtained during the in-situ biostimulation experiment. Sucrose concentrations were measured with a spectrometric method (enzymatic test kit for saccharose / D-glucose, Böhringer) from acidified samples (1 % v/v HCl suprapure, Merck). Fatty acids (acetate, butyrate, and propionate) were measured from filtered and acidified samples (4 % v/v H_3PO_4 suprapure, Merck) and analysed with a gas chromatograph (GC; model 437, Chrompack) according to GALLERT & WINTER (1997). Total plate count (TPC) representing the presence of free germs in groundwater was determined from untreated samples (in dependence on DIN 38411, TPC assignment in water) as an applicable method for the monitoring of relative changes in microbial populations (KIEFT et al. 2007).

Quality of analytical measurements. Ion balances represent as sum parameter the quality of all cation and anion measurements. They were calculated for samples taken before and during the field experiments in December 2009 with PHREEQC (see below). Deviations for samples from the high As site vary within a range of +1.23 and -15.3 (median: -9.57 %). In case of the low As site, balances are worse (median: -10.4 %) and deviations strongly increase after addition of sucrose, especially in wells B and E (up to +41.4 %), where intense microbial reactions developed. Cations were determined by HR-ICP-MS, where measurements showed a very good accuracy regarding sample replicates and reference solutions. The same applies to anion determinations by IC. Since HCO_3^- represents the principal anion (>85 %), detection inaccuracies contribute to the ion balance nearly one to one. Thus, differences are considered to result from the titrimetric determination of the TA in field. Another possibility is that not all ions were determined, especially organic compounds. This became apparent during the sucrose experiment. Here, large amounts of organic acids occurred, which were not included in the balance calculations.

Due to the comparatively limited spectrum of analysed parameters, no ion balances could be calculated for samples of the field survey and the regular monitoring samples. Comparisons of the monitoring data provided a reliable tool to identify potential outliers, since the groundwater composition of most wells remained stable until the in-situ experiments were conducted. Further considerations regarding data quality are included in chapters 6 and 7, respectively.

Hydrochemical calculations. Results of hydrochemical analysis were used to derive further important parameters. Ion balances were calculated with PHREEQC (version 2.16, thermodynamic data from the wateq-4 data base), which offers "a wide variety of low-temperature aqueous geo-chemical calculations" based on an "ion-association aqueous model" and "thermodynamic controlled water-sediment equilibrium reactions" (PARKHURST & APPELO 1999).

PHREEQC was further used to determine:

- Thermodynamic predominating element species;
- pH-E_H-stability diagrams;
- Saturation indices (SI) of mineral phases.

Saturation indices are important to estimate the behaviour of a pure phase in an aquatic environment. They are calculated from the log K_d value (distribution coefficient aqueous/solid phase) of the chemical equation (describing precipitation, respectively dissolution), as well as respective solute concentrations and their activities. A SI of 0 reflects state of equilibrium, positive SI indicate supersaturation and negative under-saturation. For example, the SI of dolomite is based on Ca, Mg and HCO_3^- concentrations in groundwater and the prevailing pH and temperature (PARKHURST & APPELO 1999). It is further mandatory to consider the prevailing redox potential, which has a decisive influence on the solubility of different metal minerals, especially Fe-(oxyhydr)oxides. Reaction rates of respective precipitation or dissolution reactions need to be estimated from concentration changes in a certain time span and space. Hence, it is very difficult to assess how fast a supersaturated mineral phase will precipitate from a certain solution. Depending on the prevailing conditions, precipitation or dissolution of calcite is for example considered a relatively fast process that can proceed within hours to months (MERKEL & PLANER-FRIEDRICH 2008). When certain parameters like pH or the CO_2 partial pressure [$p(CO_2)$] change (e.g., through microbial respiration reactions), the water-sediment equilibrium (and therefore the SI) of carbonates will accordingly adjust.

Additionally, the attempt was made to use inverse modelling to describe the mixing of different groundwater layers during the pumping experiment, but the equation system was over-determined by the high amount of included parameters.

4.4   STATISTICICAL METHODS

Statistical methods were used to support interpretation of the comprehensive sediment and groundwater data. Therefore, average values (arithmetic mean), σ and %RSD were calculated. To describe the dependence between two variables X and Y, the Pearson correlation coefficient (r_{x-y}) was calculated from outlier cleaned (Grubbs test) results, presuming a normal (Gaussian) distribution (SCHUENENMEYER & DREW 2011). In some graphics, the R^2 value (coefficient of determination) is included to describe the agreement of a regression line with respective data points. To reduce the influence of extreme values and potential outliers, median and 25 % and 75 % quartiles were calculated as well. These values were partly summarised in form of Box-Whisker-Plots (STATISTICA, StatSoft®, version 8), comprising extreme values, outliers and non-outlier boundaries, the so called whiskers (defined as 1.5* of the interquartile range between the 75 % and 25 % quartiles).

Dendrograms represent graphical summaries that were used to describe results of the sediment analysis (SCHUENENMEYER & DREW 2011). Here, variables were grouped in different clusters in dependency of their statistical distances with STATISTICA. To compensate difference in scale, data was previously z-normalised (STOYAN et al. 1997). Clusters were calculated from data with an agglomerative hierarchical cluster analysis. Euclidean distances were used to compute the geometric distance in the multidimensional space, while clustering was done by the Ward's method. This method uses the variance to evaluate the distance between presumed clusters, and identifies the best fitting clusters by iteratively minimizing the sum of squares. This step is repeated until hierarchical clusters with minimum distances are determined (WARD 1963).

5. FIELD SURVEY AND STUDY SITE SELECTION

5.1 INTRODUCTION

The following considerations focus on an aggregated data set, which was used to characterise groundwater in the investigation area. Additionally, the results formed the base to select two appropriate study sites with contrasting As groundwater concentrations (referred to as *high As site* and *low As site, see* chapter 5.3.2 and figure 5.1).

5.2 RESULTS

In contrast to the comparatively small size of the investigation area, hydrochemical properties in groundwater samples vary spatially in a wide range, especially in respect of As, Fe and PO_4^{3-} concentrations (Table 5.1). Local groundwater generally belongs to the $Ca-(Mg)-HCO_3^-$ type and redox parameters reflect predominantly reducing conditions in related aquifer parts. Vertical and horizontal distributions of dissolved As are inhomogeneous. Concentrations vary within a few decimetres (laterally) and metres (vertically) by up to two orders of magnitude (Figure 5.2). The median of As concentrations is 33.9 µg L^{-1}, while concentrations reach up to of 333 µg L^{-1} (Table 5.1). The provisional WHO (2003) drinking water guideline standard (10 µg As L^{-1}) is exceeded in 127 out of 174 samples (73 %), whereas the national Indian drinking water standard (50 µg As L^{-1}) is exceeded in half as many samples (36 %). Some of the sampled wells are recently installed governmental wells tapping water from deeper than 100 m. After realising the As problem, the regional government set up the installation of deep wells and arsenic-removing filter systems to provide a central supply with arsenic-free drinking water to the local communities.

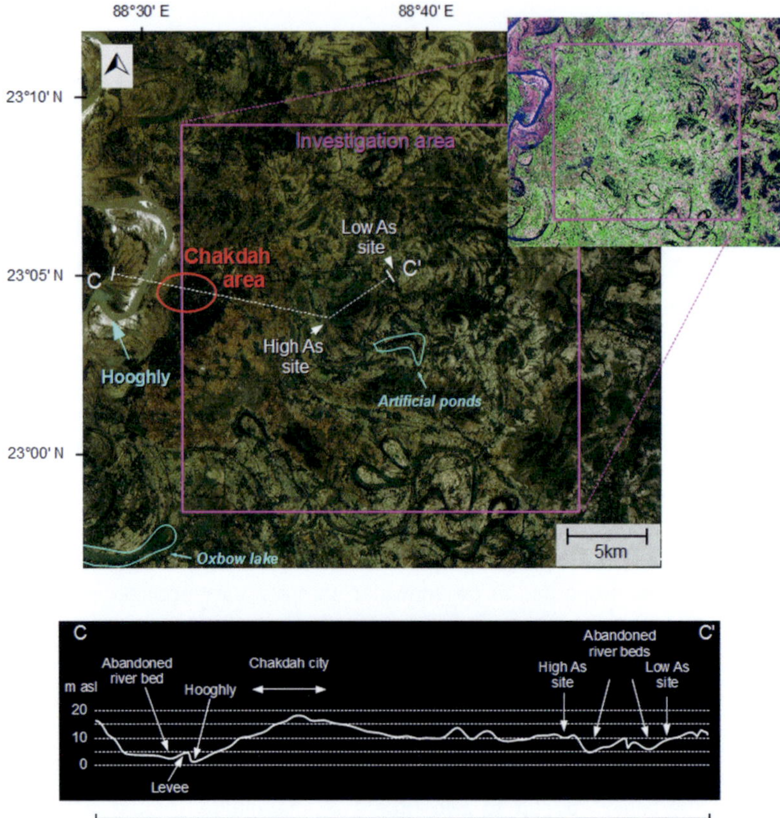

Figure 5.1: Landsat 7 imaginary and pseudo colour picture of the investigation area, including a S-W transect from the Hooghly River towards the study sites. The surface reflects an inhomogeneous and complex distribution of meanders, abandoned river beds, oxbow lakes, fields and artificial ponds. The relief profile underlines the characteristic topography of the floodplain with marginal differences in elevation. Figure caption continued on next page.

Figure 5.1, continued: Satellite image and pseudo colour picture were created with NASA World Wind 1.4, based on Landsat 7 image data. The pseudo colour picture was provided by MDA Federal Information systems for NASA World Wind. The elevation profile is super-elevated and was generated with NASA World Wind Terrain Profiler, based on the SRTM-30 PLUS digital elevation model. Resolution of satellite image: 15 m, resolution of digital elevation model: 3 arc-seconds.

According to well depths and characteristic hydrochemical compositions, groundwater samples were further separated into three different classes (Table 5.2). The deeper the wells, the higher the Na/Cl ratios and the lower the SO_4^{2-} concentrations. In wells with depths between 20 to 40 m, highest As, Fe and PO_4^{3-} concentrations occur. In range of these wells, lowest As/PO_4^{3-} mol ratios prevail. The common drilling technique (a labour-intensive variant of rotary drilling) used for installation of local private wells is strongly depth limited, which is why most wells only reach to a depth of about 24 m. Considering the complete data set, weak positive correlations exist between dissolved As, PO_4^{3-} and Fe concentrations, whilst other parameters do not show any correlations with As (Table 5.1).

Samples were further distinguished based on the NO_3^-, Mn, Fe and SO_4^{2-} concentrations, from which the prevailing redox state was derived following to the classification scheme of JURGENS et al. (2009). It is presumed that groundwater is generally suboxic to anoxic (O_2 <0.5 mg L^{-1}), and dissolved Fe and Mn are present in reduced form, as it was reported from the Chakdah area (Figure 5.3; CHARLET et al. 2003 & 2007, MÉTRAL et al. 2008). Based on these results, two locations with contrasting As concentrations in shallow groundwater were chosen as high As site and low As site (Figure 5.2). An overview of the groundwater compositions at both sites is given in Table 5.3, while the decision to choose these locations is discussed in chapter 5.3.2.

Table 5.1: Summary of the field survey results comprising respective median, lower and upper quartiles (25 % and 75 % Q.) to express characteristic concentration ranges of solutes in samples from different depths (n: 174).

Value range	Depth (m bls)	pH	EC (μS/cm)	Ca (mg/L)	Mg (mg/L)	Na (mg/L)	K (mg/L)	Cl$^-$ (mg/L)
Minimum	7	6.5	200	20.6	5.42	4.41	0.77	0.67
25 % Q.	21	6.8	591	81.6	18.4	13.0	2.40	6.29
Median	24	7.1	668	96.3	21.7	17.2	3.01	12.3
75 % Q.	36	7.4	768	115	25.6	26.6	3.74	25.6
Maximum	270	8.0	1514	177	43.7	112	44.5	147
Average	40	7.1	696	98.7	22.5	21.7	3.73	21.1
r_{As-}	-0.02	-0.12	+0.03	+0.00	+0.09	+0.09	-0.10	-0.16

Value range	DOC (mg/L)	NO$_3^-$ (mg/L)	SO$_4^{2-}$ (mg/L)	PO$_4^{3-}$ (mg/L)	Mn (mg/L)	Fe (mg/L)	As (μg/L)
Minimum	0.12	<0.88	<0.85	0.01	0.01	0.01	0.31
25 % Q.	1.38	<0.88	<0.85	0.18	0.16	0.47	7.99
Median	1.76	1.23	<0.85	1.29	0.27	2.47	33.9
75 % Q.	2.50	2.18	7.08	3.63	0.40	4.55	69.4
Maximum	20.5	550	50.3	10.1	2.53	45.9	333
Average	2.16	5.84	5.20	1.99	0.32	3.72	51.7
r_{As-}	+0.04	-0.06	-0.29	+0.53	+0.02	+0.48	-

Table 5.2: Summary of selected parameters that express characteristic values for groundwater samples from different depth intervals (n: 165, nine samples with unknown depths were removed).

Depth	Range	pH	Ca (mg/L)	Mg (mg/L)	Na (mg/L)	Cl (mg/L)	NO_3^- (mg/L)	Na/Cl* (mol ratio)
<20 m (n: 37)	25 % Q.	6.9	94.3	18.6	13.6	10.7	<0.88	1.01
	Median	7.2	102	22.4	18.7	19.3	1.21	1.35
	75 % Q.	7.4	122	26.0	23.5	33.7	1.98	2.28
20-40 m (n: 92)	25 % Q.	6.8	80.3	17.5	13.1	7.79	<0.88	1.12
	Median	7.1	101	21.7	17.8	15.8	1.40	1.66
	75 % Q.	7.4	118	26.0	26.8	32.3	2.16	2.91
>40 m (n: 36)	25 % Q.	6.9	73.4	20.0	12.8	2.52	<0.88	4.77
	Median	7.1	85.1	21.6	16.1	3.78	<0.88	7.36
	75 % Q.	7.4	95.2	23.1	27.4	5.39	1.79	11.4

Depth	Range	SO_4^{2-} (mg/L)	PO_4^{3-} (mg/L)	Mn (mg/L)	Fe (mg/L)	As (µg/L)	As/PO_4^{3-} (mol ratio)
<20 m (n: 37)	25 % Q.	<0.85	0.11	0.21	0.18	3.10	0.01
	Median	6.83	0.43	0.33	1.71	17.0	0.03
	75 % Q.	13.4	1.74	0.48	3.39	38.4	0.06
20-40 m (n: 92)	25 % Q.	<0.85	1.21	0.20	2.25	21.1	0.01
	Median	<0.85	3.06	0.31	3.81	46.2	0.02
	75 % Q.	6.14	4.23	0.39	6.24	88.6	0.04
>40 m (n: 36)	25 % Q.	<0.85	0.06	0.08	0.05	1.49	0.02
	Median	<0.85	0.14	0.14	0.46	14.9	0.15
	75 % Q.	<0.85	0.46	0.28	1.52	82.2	0.26

*Na/Cl mol ratio determined from local rain water: 0.41

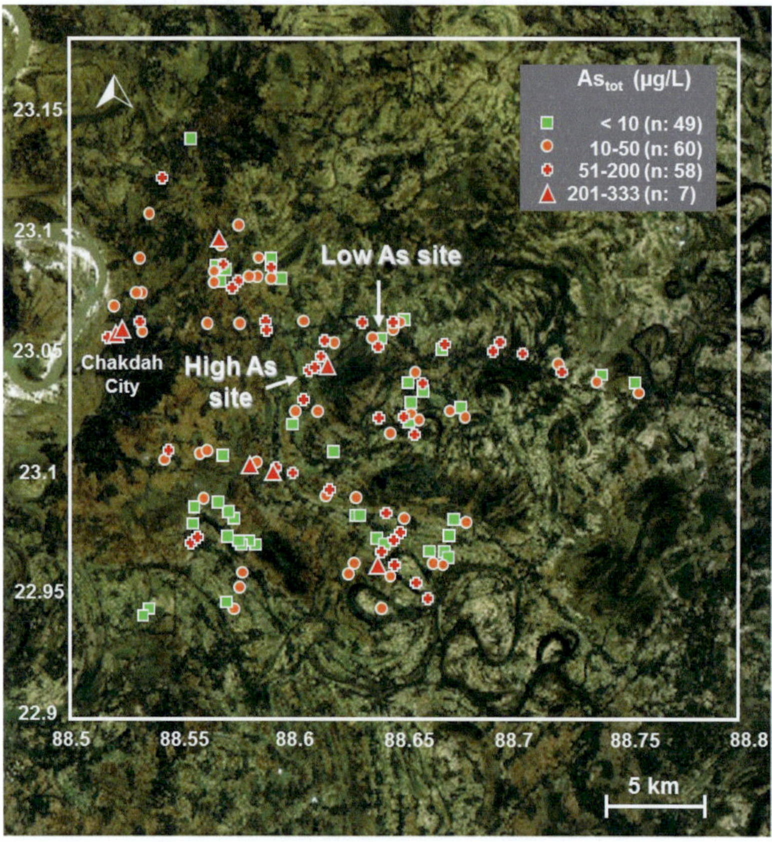

Figure 5.2: Location and As concentrations of sampled tube wells (n: 174) within the investigation area, east of Chakdah and the Hooghly River. Samples are grouped into four classes according to the As concentrations:

1) < 10 µg L^{-1}, the WHO threshold value;

2) > 10 µg L^{-1} and < 50 µg L^{-1}, the Indian threshold value;

3) > 50 µg L^{-1} and < 200 µg L^{-1};

4) > 200 µg L^{-1} and < 334 µg L^{-1}.

Underlying Landsat 7 imaginary created with NASA World Wind 1.4 (resolution: 15 m). Arsenic distribution plotted with Surfer (version 7, Golden Software).

Table 5.3: Summary of the groundwater hydrochemistry at the high As site and the low As site. At the high As site, four other wells were sampled within ~500 m around well 132, located in the villages Sahispur and Maturagachi. At the low As site, 8 surrounding wells were sampled in the villages Chakudanga and Amdangaround, all within ~700 m around well 125. Well 122 represents a public well equipped with an As filter system, with raw groundwater (122a) and filtered water (122b). Well 120 is located at a primary school.

Well	Depth (m bls)	pH	Na (mg/L)	K (mg/L)	Ca (mg/L)	Mg (mg/L)	Cl^- (mg/L)	NO_3^- (mg/L)	SO_4^{2-} (mg/L)	PO_4^{3-} (mg/L)	DOC (mg/L)	Mn (mg/L)	Fe (mg/L)	As (µg/L)
							High As site							
132	24	7.1	26.8	3.33	91.7	22.5	18.5	1.40	1.44	4.23	1.47	0.49	3.30	285
131	24	7.3	18.5	1.78	77.0	17.5	10.9	<0.88	<0.85	3.74	0.99	0.51	5.46	84.9
179	55	7.1	17.0	2.32	66.9	19.0	6.67	<0.88	3.99	0.70	1.54	0.16	0.03	120
180	24	7.2	25.8	1.63	74.3	19.8	7.04	1.60	<0.85	4.90	1.56	0.11	3.88	116
181	20	7.4	14.0	3.03	71.6	17.3	4.23	1.34	<0.85	4.77	1.22	0.21	3.65	124
							Low As site							
125	24	7.2	15.1	4.03	109	24.1	46.7	1.68	10.7	0.03	1.07	0.51	0.05	2.29
117	30	7.1	15.6	2.20	83.5	20.9	21.9	1.72	16.3	4.16	1.24	0.37	6.54	37.9
118	?	7.0	9.65	3.71	105	20.2	14.3	1.17	3.91	0.57	6.24	0.57	13.9	4.62
119	18	7.3	9.29	2.65	113	16.4	10.7	1.43	4.51	0.14	2.78	0.40	1.38	3.24
120	183	7.2	14.3	3.42	108	21.2	4.96	1.88	<0.85	0.07	2.04	0.06	1.30	34.8
121	25	7.2	10.8	2.72	80.8	20.9	10.9	<0.88	2.00	3.79	0.89	0.18	4.05	50.4
122a	167	7.5	9.46	2.36	71.0	14.0	2.25	<0.88	<0.85	0.41	1.34	0.06	1.07	77.4
122b	167	7.3	13.2	3.30	94.2	19.4	2.31	3.13	<0.85	0.50	1.86	0.08	1.28	101
124	24	7.3	13.8	3.69	85.9	16.4	3.55	<0.88	<0.85	2.58	2.30	0.47	4.85	160
126	24	7.2	10.8	3.01	71.3	14.3	3.56	1.38	5.42	3.83	0.74	0.27	1.04	61.0

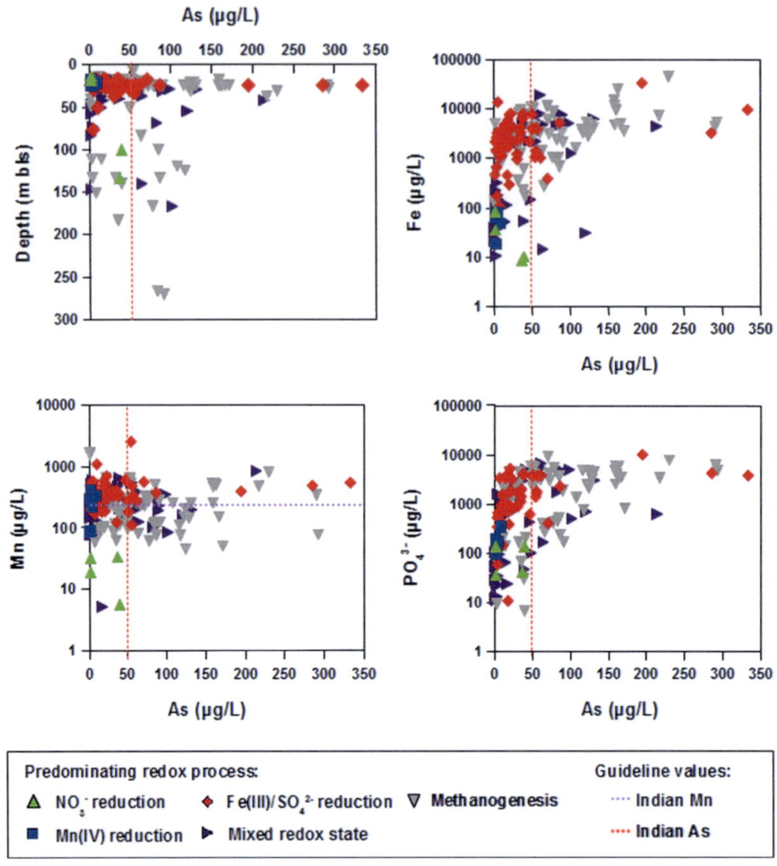

Predominating redox process	NO$_3^-$ (mg/L)	Mn (mg/L)	Fe (mg/L)	SO$_4^{2-}$ (mg/L)	n (avg. As µg/L)
NO$_3^-$ reduction	≥ 2.215	< 0.05	< 0.10	-	4 (19.4)
Mn(IV) reduction	< 2.215	≥ 0.05	< 0.10	-	5 (2.89)
Fe(III)/SO$_4^{2-}$ reduction	< 2.215	> 0.05	≥ 0.10	≥ 0.90	53 (38.3)
Methanogenesis	< 2.215	> 0.05	≥ 0.10	< 0.90	62 (82.2)
Mixed redox state					50 (36.7)

Figure 5.3: Profile of As distributions plotted against well depths. Results display that highest As concentrations appear between 20 to 40 m depth. Figure caption continued on next page.

Figure 5.3, continued: Plots of As-Fe, As-Mn and As-PO$_4^{3-}$ concentrations emphasise close correlations between these solutes and the predominating redox state. Number of samples included: 174. The redox state classification following JURGENS et al. (2009) bases on the given threshold values. Due to the SO$_4^{2-}$ quantification limit (lq) of 0.85 mg L^{-1}, the respective threshold was increased from 0.50 to 0.90 mg L^{-1}.

5.3   DISCUSSION

5.3.1   GEOCHEMICAL CHARACTERISATION OF LOCAL GROUNDWATER

Redox processes. In the following, presented results of the hydro-chemical survey are used to answer the questions raised in the introduction (chapter 1.2). The flat topography of the Bengal Basin and constantly slow horizontal groundwater flow velocities entail intensive water-rock interactions (MICHAEL & VOSS 2009b). Results of groundwater dating conducted in nearby areas revealed that shallow groundwater in Holocene aquifer sediments is not older than 100 years (HARVEY et al. 2005, MC ARTHUR et al. 2010, STUTE et al. 2007). Increase and/or decrease of redox sensitive parameters reflect decisive influences of anaerobic microbial metabolic processes, of which different TEA consuming redox processes were identified, ranging from NO$_3^-$ reduction over Mn(IV), Fe(III) and SO$_4^{2-}$ reduction to methanogenesis.

The applied redox classification concept after JURGENS et al. (2009) is generally a simplification and does not necessarily match with situations met in complex natural systems, where groundwater is often not in redox equilibrium. Here, multiple redox reactions may partly overlap at a certain point of time, although usually one process dominates (JURGENS et al. 2009, STUMM & MORGAN 1996). For example, NO$_3^-$ and Mn(IV) reduction often occur parallel to each other, as well as Fe(III) and SO$_4^{2-}$ reduction. Results reflect that groundwater is mainly in state of Fe(III)/SO$_4^{2-}$ reduction, methanogenesis, or in a mixed redox state (Figure 5.3). It was not possible to further differentiate between $Fe(III)/SO_4^{2-}$ reduction since no sulphide determination was done (MC MAHON & CHAPELLE 2008). Spatial variations of these redox indicators reflect the presence of inhomogeneous redox zones in the aquifer. This aspect is discussed later in detail for the two study sites (chapters 6.3.4 and 7.3.4).

Indications for arsenic release. The hydrochemistry of groundwater from the investigation area supports the assumption that microbially mediated processes are the key mechanism for As mobilisation. The presence of As concentrations exceeding 50 µg L^{-1} is limited to samples in state of $Fe(III)/SO_4^{2-}$ reduction, methanogenesis or a mixed redox state (see Figure 5.3). The more reducing the redox conditions, the higher the concentrations of dissolved As. This trend manifests in a moderate positive correlation among As and Fe concentrations (correlation coefficient: +0.46), pointing at release according to the Fe(III) reduction hypothesis (chapter 2.2.3). It is noteworthy that samples in state of $Fe(III)/SO_4^{2-}$ reduction were found to have not yet reached As concentrations found in samples that are in state of methanogenesis. This can be explained by the assumption that these samples originate from more mature aquifer parts, where Fe(III) reduction and associated As release is completed or at least non-dominant.

Samples indicating mixed redox states. A considerable number (50) of samples did not fit into one of the four principal redox classes and were therefore assigned as "mixed redox state". Two important types can be distinguished within these samples. The first group includes 34 samples with elevated As concentrations, where the state of $Fe(III)/SO_4^{2-}$ reduction or methanogenesis should be reached, but NO_3^- concentrations clearly exceed the threshold value of 2.25 mg L^{-1}. This is attributed to mixing processes, which are caused by infiltration of nitrate-rich surface water into more reducing zones. For example, extensive pumping has been proven to cause such effects in case of the Chakdah area (CHARLET et al. 2007), as well as in Hanoi, Vietnam (NORRMAN et al. 2008).

The second group contains 16 samples that appear to be in a transition between Mn(IV) and Fe(III)/SO_4^{2-} reduction. These samples hold low NO_3^- and Fe concentrations that remain closely below the threshold values of the classification scheme, while Mn concentrations are increased. Here, measured As concentrations are comparatively low, except for three samples. Results point at precipitation and adsorption processes, which are discussed in the following.

Arsenic and phosphate. The strongest correlation between As concentrations and any other determined parameter in all samples is related to PO_4^{3-} (r_{As-P}: +0.53). Phosphate in groundwater generally originates from mineralisation of organic matter and from reductive dissolution of phosphate-hosting Mn- and Fe-(oxyhydr)oxides, which is known from oxbow lake sediments (LEWANDOWSKI & NÜTZMANN 2010, O'DAY 2006). Phosphate is further considered a strong competitive anion that can induce As release from host surfaces. Since As(III/V) and PO_4^{3-} have similar geochemical properties and preferentially adsorb to surfaces of Fe-(oxyhydr)oxides (GOH & LIM 2004), it is here more likely that both are controlled by the same mobilisation mechanism, which would be the activity of FeRB. Additionally, P could be fertiliser derived and enter the aquifer via infiltrating irrigation water (ACHARYYA et al. 2000).

Highest concentrations of dissolved As, PO_4^{3-} and Fe occur in wells that are 20 to 40 m deep. Only a few samples (n: 36) were taken from wells exceeding 40 m in depth, but this distribution is in accordance to results from other surveys that have been previously conducted in the Bengal Basin (BGS & DPHE 2001, HARVEY et al. 2002). Here, dissolved As, PO_4^{3-} and Fe concentrations never were that high, or enrichment processes have stopped and solutes were slowly, but constantly flushed-out into the Bay of Bengal. In the BDP, increasing depth is equivalent to increasing groundwater age (HARVEY et al. 2005, MC ARTHUR et al. 2010). This increases the influence of kinetic based solid-water equilibrium reactions like the precipitation of supersaturated mineral phases. Hence, precipitation of supersaturated minerals, especially Fe-minerals, is another potential mechanism able to explain this depth depending decline of As and PO_4^{3-}

concentrations in groundwater. Precipitation of Fe-minerals enables an immobilisation of dissolved As and PO_4^{3-}. This aspect is discussed later on in the discussion of the low As site (chapter 6.3.4) and in the final summary (chapter 8.2).

Manganese. In addition to the problematic As concentrations, manganese exceeds in 79 samples the Indian drinking water threshold value of 0.30 mg L^{-1} (if no other drinking water source is available, otherwise 0.10 mg L^{-1}; IS 10500, reaffirmed 1993). This is known from other areas of the BDP as well and is considered to originate from the reductive dissolution of Mn-minerals under mildly reducing conditions (MC ARTHUR et al. 2012, VAN GEEN et al. 2007). The occurrence of neurobiological effects related to chronic Mn uptake is still under debate, but the WHO decided to remove the provisional guideline recommendation in the latest edition of the guideline values for drinking water (WHO 2011). This was decided because the former guideline value of 0.40 mg L^{-1} was considered well above concentrations normally found in drinking water, which is definitively not true in case of the BDP (MC ARTHUR et al. 2012).

Arsenic in public wells. Most people meanwhile rely on official governmental wells, which deliver water from deep aquifer parts and are partly equipped with As filters (Figure 5.4). Pipelines were recently installed to provide a central water supply with treated surface water. Nevertheless, some small communities still rely on private tube wells. All in all, 15 wells turned out to be critically burdened with As in the investigation area, reaching high concentrations of 159 to 333 µg As L^{-1}. Unfortunately, providing technical solutions exceeded the competence of the project. The Indian project partner intends to contact the corresponding local responsibilities in order to find a solution for critically burdened wells with As concentrations well above the national Indian threshold value, including one public school well (GSFC school, 83.0 µg As L^{-1}).

Figure 5.4: Picture of a public well installed in 2003 in the village Chakudanga, which is equipped with an As filter cartridge (well depth: 163 m). Surprisingly, the As concentration in filtered water was higher as compared to the raw water (77.4 compared to 101 µg L^{-1}), strongly suggesting that the filter material needs to be replaced. This example demonstrates that current mitigation strategies (use of deep wells and filter systems) can be unreliable.

5.3.2 STUDY SITE SELECTION

Based on the results of this survey and considering the proposed monitoring and field experiments, two appropriate study sites were selected. The underlying idea of this concept is to compare two sites that are located within the same area, but hold contrasting As concentrations in shallow groundwater. With this approach information regarding the As release mechanism shall be derived.

To find two capable sites that are characterised by high, respectively low dissolved As concentrations, results of the field survey were analysed regarding manifold selection criteria.

"Hard" selection criteria for each site:

- Adequate (high/low) concentrations of dissolved As in regard to the WHO threshold value of 10 µg L^{-1} and characteristic local contents;

- Reducing redox state (indicated by NO_3^-, Mn, Fe and SO_4^2 concentrations) to prove if As is released during Fe(III) reduction;

- Restriction to shallow groundwater (maximum depth 40 m), which is the preferred zone of As enrichment as well as the predominating depth of local tube wells;

- Adequate distance to the well investigated Chakdah area to avoid influences of the local depression cone underneath the city.

"Weak" selection criteria for each site:

- Easy accessibility by car and small distances between the candidate sites to facilitate the regular monitoring sampling;

- Low concentrations of compounds toxic to microbes (Cu and Cd);

- Possibility to discharge large volumes of water during pumping;

- A minimum distance between the Hooghly River and the candidate sites to avoid influences of flooding on the hydrology.

According to results of the field survey, the small village Sahispur is considered to be located above an As plume. Concentrations of As in the groundwater sample from well 132 reaches 285 µg L^{-1}, while four nearby wells hold concentrations of 85.0 to 124 µg L^{-1} in depths between 20 to 55 m (Table 5.3). Groundwater of these wells is typically enriched in dissolved Ca, Fe, Mn and PO_4^{3-}, with low NO_3^- and SO_4^{2-} concentrations.

Hence, results indicate moderate to strong reducing conditions in the shallow aquifer. The location around well 132 meets the criteria mentioned above and was therefore chosen as "high As site", located at 23°04'15.5"N and 88°36'33.5"E (Figure 5.2). Here, five nesting monitoring wells were installed. The monitoring well setup is shown in Figure 5.5.

In contrast, water from well 125 was measured with only 2.29 µg As L^{-1}. This well is located at the rim of the larger village Chakudanga (around 2,000 inhabitants) in a distance of about 3.1 km to the high As site. Well 125 is further situated next to an unpaved cricket field and a young tree plantation, providing a suitable environment for the in-situ biostimulation experiment. The hydrochemistry turned out to be less reducing than at the low As site, holding 10.7 mg L^{-1} SO_4^{2-} and 0.05 mg L^{-1} Fe. In the surrounding of well 125, 8 additional wells were sampled. Results display highly variable concentrations of major and trace elements like Cl^-, Fe, and As in the shallow groundwater (Table 5.3). Since local conditions at well 125 were considered appropriate, this area was chosen as low As site (coordinates: 23°04'58.2"N, 88°38'13.1"E, Figure 5.2). Again, five nested monitoring wells were installed similar to the high As site.

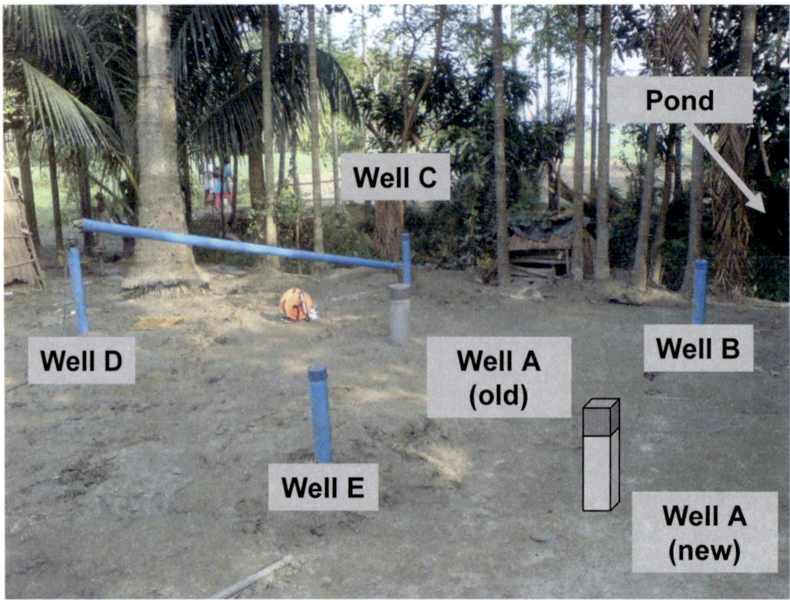

Figure 5.5: The monitoring well setup of the high As site, directly before the concrete sealing was placed. Included is the position of the new central well A, which was installed after the old one collapsed soon after installation.

6. THE LOW ARSENIC STUDY SITE

6.1   INTRODUCTION

The low As site was chosen to investigate the influence of the availability of OM on the release of As under in-situ field conditions (NEIDHARDT et al. 2012b). In contrast to extensive research on the spatial distribution of As in groundwaters of the BDP, very few systematic field experiments have been carried out in order to reveal the relationships between geochemical reactions and microbial mechanisms that mobilise As from aquifer sediments and to verify preliminary findings (HARVEY et al. 2002, SAUNDERS et al. 2008). Various methods originating from different fields of Environmental Science (Geochemistry, Hydrogeology and Environmental Microbiology) were combined to cover the most probably influences related to As release and enrichment in the scope of this work.

Investigations involved a unique and interdisciplinary approach, which combined comprehensive laboratory analyses, regular groundwater monitoring and an in-situ field experiment in four different phases:

- Biogeochemical analysis of autochthonous aquifer sediments was used to determine As host phases and characterise available OM. Microbial incubation experiments using aquifer sediments from the two study sites were conducted at the IBA to estimate the As release potential and to prepare the in-situ field experiment.

- A one year long hydrochemical monitoring of groundwater was used to capture the status quo of the aquifer. Therefore, five adjacent nested multilevel wells were installed at the study site and sampled in regular intervals of two weeks.

- An in-situ biostimulation experiment was conducted, introducing organic carbon (sucrose) into the local aquifer. Influences on the groundwater were recorded during the following three weeks through an elaborate hydro-chemical monitoring.

- Potential mid-term effects of this experiment on the groundwater composition were monitored during the following eight months.

6.2 RESULTS AND INTERPRETATION

6.2.1 SEDIMENT CHARACTERISATION

Stratigraphy and geochemistry. Well screens of the five monitoring wells are situated within a single, shallow aquifer. The sandy aquifer sediments are covered by olive-brown, thin layers that reach down to 3.85 m bls and comprise varying proportions of clay and silt (referred to as facies F4, Figure 6.1). The top 0.60 m are compacted and contain fragments of clinker. Between 1.90 and 3.85 m bls, mollusc shells as well as reddish and black mottles of Mn- and Fe-(oxyhydr)oxides occur, while Ca and TIC contents are here lower than in the surrounding sediments (Figure 6.2 and APPENDIX II, Table A 2.1). In this part, relatively high contents of As, Fe, Mn, Ni, Cu, Zn as well as TOC and TS appear. Between 3.20 to 3.85 m bls, a thin layer of silty fine sand is located, where the sediment colour changes into dark grey. Below, fine to medium grained sand dominates down to the target depth of 45.5 m bls. Due to the limited drilling depth, no information is available regarding the total thickness of the aquifer. From 3.85 to approximately 19 m bls (facies F3b), silt contents, TOC, major and trace elements (except for Zr and Ce) gradually decrease with depth in the silty medium and fine sands. The vertical distributions of sedimentary As, TOC and TS constitute an exception, they peak within a thin, brown coloured segment composed of fine sandy silt around 7.75 m bls. From about 19 to 29 m bls (facies F3a), silt is completely absent and previously decreasing elements stabilise at the lowest levels of the respective contents. Remaining sediments down to 45.5 m bls (facies F2) are composed of medium sandy fine sands, which hold throughout low proportions of silt and an embedded gravel lens in about 31 to 33 m bls. In this part, most element contents strongly vary similar to the upper layers of F4. Sediments from facies F1 do not occur within this profile and are described in context of the high As site (chapter 7.2.1).

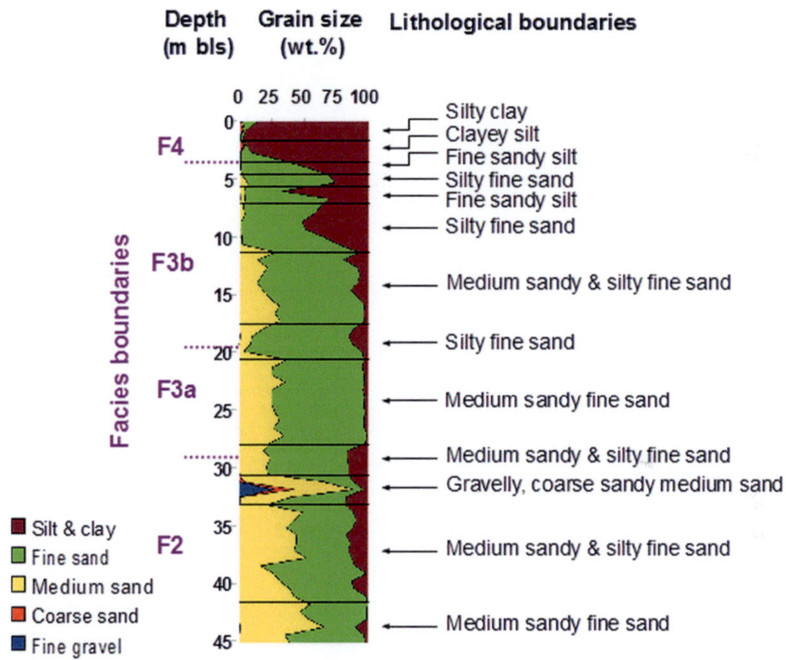

Figure 6.1: Grain size distribution of the sediment samples, including litho-facies boundaries.

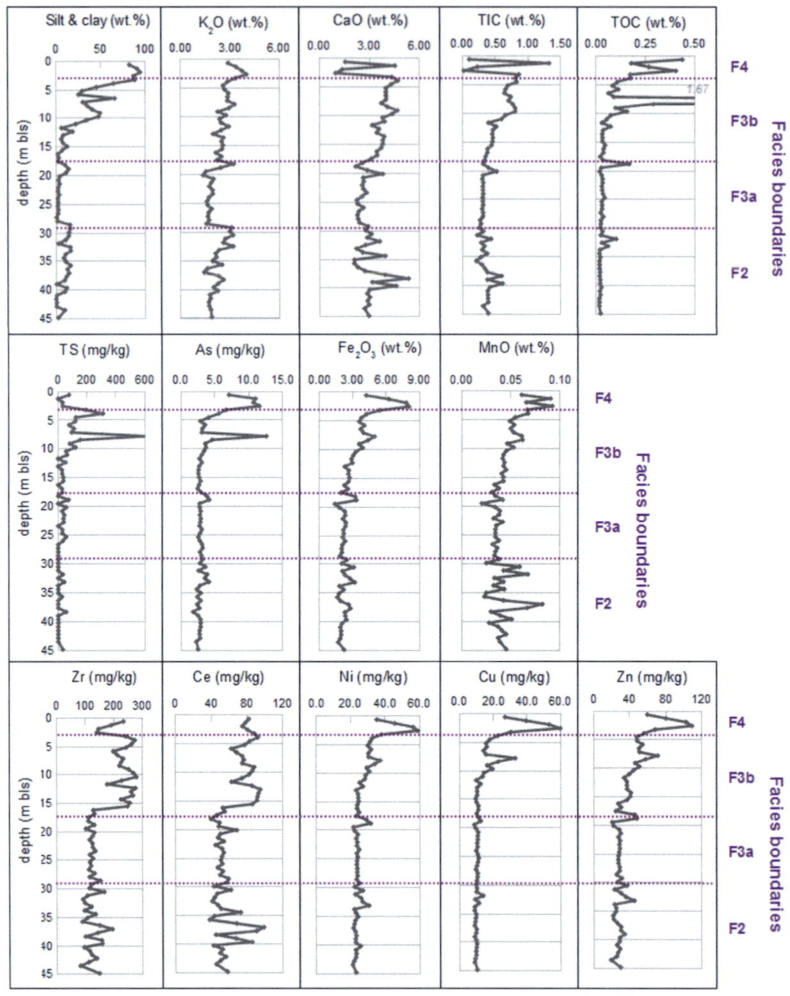

Figure 6.2: Vertical distribution of major and trace elements, TOC, TIC, TS and silt and clay contents. The profile comprises 70 samples in intervals of 0.65 m.

Mineralogy. Mineralogical constituents are summarised in Table 6.1. Detrital quartz grains represent the greater part of the aquifer sediments, followed in frequency by feldspar (according to XRD spectra presumably anorthite, see APPENDIX II, Figure A 2.1), carbonate (calcite and dolomite), mica (muscovite) and chlorite (clinochlore). All samples from the clayey surface sediments (facies F4) were analysed for the respective clay mineral inventory. Results reveal the presence of smectite and potentially traces of kaolinite and illite (see APPENDIX II, Figure A 2.2), which are considered common constituents of BDP sediments (MUKHERJEE et al. 2009).

Pre-concentration of the sample material by magnetic separation allowed identification of iron-rich minerals from concentrates. Traces of Fe-oxides (magnetite, hematite), garnet (almandine), biotite (phlogopite, which could be separated from muscovite), chloritoid, actinolite and presumably epidote could be additionally distinguished in respective concentrate fractions (APPENDIX II, Figures A 2.3 and A 2.4).

Table 6.1: Summary of identified minerals in representative sediment samples. With "?" marked minerals could not be proved beyond any doubt due to weak and partly superimposed XRD peaks.

Main mineral phases		
• Quartz	• Dolomite	• Smectite
• Feldspar	• Muscovite	• Kaolinite (?)
• Calcite	• Chlorite (clinochlore)	• Illite (?)
Minor mineral phases		
• Magnetite	• Actinolite	• Fe(III)-oxides (hematite ?)
• Chloritoid	• Biotite (phlogopite)	• Amorphous Fe-(oxyhydr)oxides
• Almandine	• Epidote (?)	

Arsenic content. Despite the comparatively high arsenic contents in samples of the clayey surface layers, bulk As contents in the aquifer sediments range between 1.8 and 12.6 mg kg^{-1} (average: 3.2 ± 1.3 mg kg^{-1}; n: 64). The depth distribution of As correlates significantly (r >+0.75) with Fe, TOC, the fine grain fraction (<0.063 mm) and trace elements that are typically associated with iron bearing minerals (e.g., Ni, Cu and Zn) (Table 6.2). The comprehensive results are summarised in form of an agglomerative hierarchical cluster analysis, which groups the different variables (analytical parameters) in form of a dendrogram in dependence of their statistical similarities (Figure 6.3). This dendrogram indicates a close statistical distance of As to Fe and affiliated trace elements (Cu, Ni and Zn).

A SEP was performed to obtain detailed information about the sedimentary arsenic-bearing host phases (Figure 6.4 and APPENDIX II, Table A 2.2). Results indicate that a considerable percentage (42.4-84.7 %) of the total sedimentary As content is PO_4^{3-}-extractable (interpreted as strongly adsorbed fraction), although this fraction was found to be overestimated at the expense of fraction III (see chapter 4.2). Remaining As is associated with (a) acid volatile sulphides, carbonates, Mn-(oxyhydr)oxides and very amorphous Fe-(oxyhydr)oxides (fraction III, average: 14.5 %), (b) amorphous Fe-(oxyhydr)oxides (fraction IV, average: 6.1 %) and (c) crystalline Fe-(oxyhydr)oxides (fraction V, average: 8.6 %) as possible hosts. Only a minor fraction (average: 1.4 %) is SO_4^{2-}-extractable (fraction I, interpreted as weakly bound phase). Samples from depths below 13 m bls are characterised by similar distribution patterns and hold throughout low As contents of <3 mg kg^{-1}. In the two surface near clayey samples, As contents are higher, and As is additionally (d) associated with silicates (18.5 % in 2.60 m bls) and (e) As-sulphides, OM and/or refractory minerals (8.7 % in 3.20 m bls). These conclusions agree in general with findings from the well investigated Chakdah site (CHARLET et al. 2007, MÉTRAL et al. 2008). In contrast to other study sites in the BDP (AKAI et al. 2004, HARVEY et al. 2002, NICKSON et al. 2000), sulphides and OM as potential As sources are only in two samples detectable (Figure 6.4). According to the generally very low TOC contents in the aquifer sediments (<0.1 %), OM is considered to play a minor role as As host in general.

Table 6.2: Statistical summary of relevant element contents in the aquifer sediments (ranging from 3.85 to 45.5 m bls, n: 64). Included are median values, lower and upper quartiles (25 % and 75 % Q.), minimum and maximum contents as well as average values (arithmetic mean) and correlation coefficients.

Range	Silt & clay (wt.%)	K_2O (wt.%)	CaO (wt.%)	TOC (wt.%)	Fe_2O_3 (wt.%)	MnO (wt.%)	TS (mg/kg)	Ni (mg/kg)	Cu (mg/kg)	Zn (mg/kg)	As (mg/kg)
Minimum	0.19	1.37	2.08	0.02	1.40	0.02	<26	21.2	8.73	19.0	1.8
25 % Q.	2.71	1.82	2.63	0.02	2.05	0.03	<26	23.2	9.75	27.1	2.8
Median	9.85	2.25	2.98	0.03	2.34	0.04	26.6	24.0	10.4	29.8	2.9
75 % Q.	15.7	2.60	3.80	0.04	2.95	0.05	54.6	27.1	11.2	38.3	3.2
Maximum	66.0	3.29	5.33	1.67	4.99	0.08	610	37.4	33.1	71.3	12.6
Average	13.0	2.25	3.18	0.07	2.58	0.04	51.5	25.4	11.6	33.6	3.2
r $_{As-}$	+0.76	+0.64	-0.11	+0.75	+0.86	+0.67	+0.53	+0.89	+0.92	+0.87	-

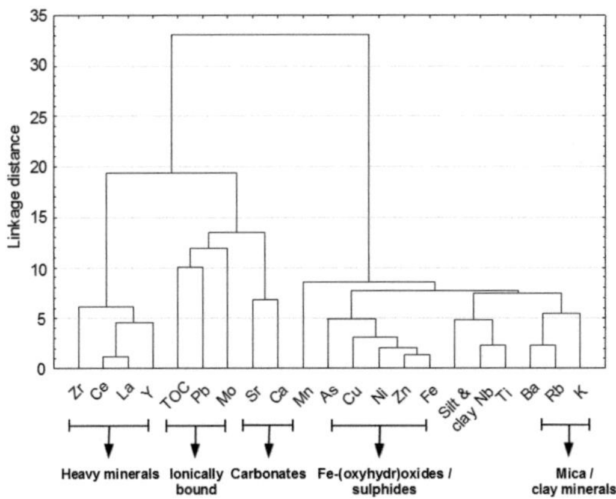

Figure 6.3: Dendrogram for 21 variables expressing the statistical distance between elements and other parameters in aquifer and aquitard sediments (0.0 to 45.5 m bls, n: 69; Ward's method and Euclidean distances; one outlier removed). Arsenic in the sediments appears to be closely associated with Fe, Zn, Ni and Cu. This group is interpreted to be primarily composed of Fe-(oxyhydr)oxides (data summarised in APPENDIX II, Table A 2.1).

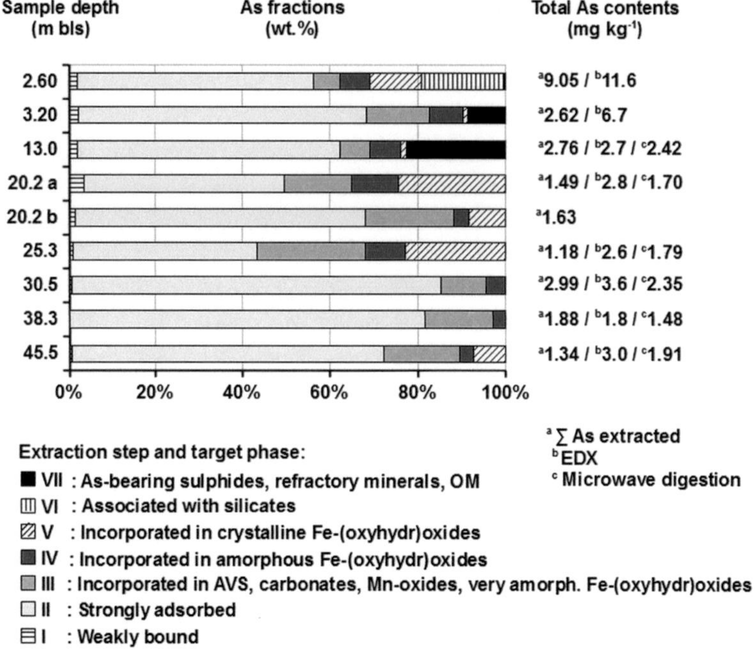

Figure 6.4: Sequential extraction procedure (according to EICHE et al. 2008) results for 9 samples that represent different parts of the aquifer. Sedimentary As is primary strongly adsorbed (extraction step II) and associated with fractions that comprise various forms of Fe-(oxyhydr)oxides (steps III-V).

Sedimentary organic matter. Since no peat layers were observed during drilling, sedimentary OM solely appears in dispersed form. In the organic-enriched clayey facies F4, TN contents as well as $\delta^{13}C$ and $\delta^{15}N$ values gradually decrease with depth, while the C/N ratios (weight ratio of TOC to TN) remain stable at values of about 8 (Figure 6.5). In sandy aquifer sediments of facies F3b beneath, C/N ratios shift to values above 10, while most $\delta^{13}C$ values vary between -23 and -20 ‰, and $\delta^{15}N$ values stabilise around + 2.5 ‰. Due to the throughout low TOC contents, no samples could be analysed from facies F3a (19–29 m bls), and only two from facies F2 (29-45 m bls).

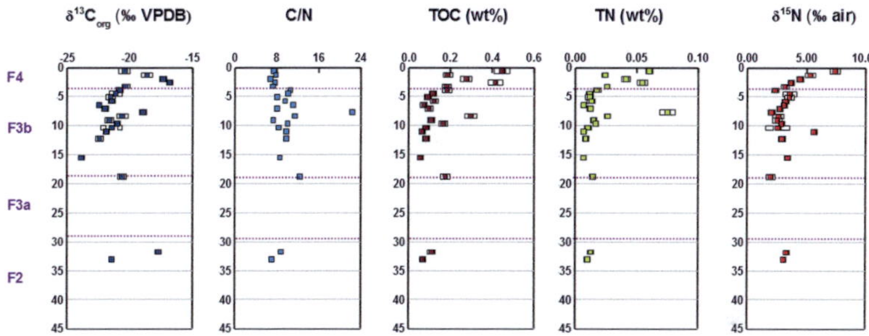

Figure 6.5: Detailed characterisation of OM from sediment samples, including $\delta^{13}C$ values, C/N-ratios, TOC fraction, TN contents and $\delta^{15}N$ values (n: 22).

The comparison of C/N-ratios, $\delta^{13}C$ and $\delta^{15}N$ values is known to serve as a reliable and commonly used biomarker for source identification in fluvio-deltaic sedimentation environments (HOEFS 2009). In addition, isotopic characterisation of OM can be included in sequence stratigraphic considerations to interpret palaeo-environmental sequences in coastal sediment deposits (LAMB et al. 2006). In the Bengal Basin, different sediment facies occur that witness tectonic activities, eustatic sea level changes and climate changes, although erosion processes and subsequent valley and channel fills have locally disrupted local facies successions (MIALL 1996). Thus, deposits can contain varying ratios of terrestrial and marine derived organic matter. The use of δ^{13}C values allows differentiating between OM derived from C_3 and C_4 land plants. By combining δ^{13}C with C/N-ratios, it is further possible to distinguish between C_3 and C_4 plant signatures and particulate organic matter (POM) derived from marine or freshwater algae and microbes (HOEFS 2009). Another common isotopic marker to distinguish marine from terrestric sources is the δ^{15}N value.

Average C/N ratios indicate that algae and microbes (freshwater and marine) as well as C_4 plants are the potential origin of the respective sedimentary OM (APPENDIX II, Table A 2.3). By comparing C/N ratios with δ^{13}C values, potential OM sources can be further differentiated (Figure 6.6). Samples from the sandy aquifer show C/N ratios of 6 to 12, and δ^{13}C values of -24 to -20 ‰ that are characteristic for OM derived from marine POM (HOEFS 2009, LAMB et al. 2006, MEYERS 1994, SARKA et al. 2009, SHARP 2006). Despite this, δ^{15}N values in this part of the aquifer are in a characteristic range of terrestric mangrove trees (C_3 plants), although most are very close the threshold value of marine POM (MUZUKA & SHUNULA 2006). A few signals even plot in range of δ^{15}N values of marine POM samples from sediments of the Bay of Bengal (GAYE-HAKE et al. 2005). Only few data is available regarding δ^{15}N values for plants in general and for mangrove plants in particular. Hence, interpretation of the δ^{13}C and C/N values that bases on a sound data base should be preferred to δ^{15}N signature interpretation. Nevertheless, influences of terrestric C_3 plant matter is typical for estuary environments and cannot be excluded in this case (MIALL 1996).

In the youngest silty and clayey deposit of facies F4, δ^{13}C values occur (> -20 ‰) that represent mixed signals, indicating an increasing influence of terrestric C_4 plant matter in the recent past. This is further supported by increasing δ^{15}N signals. Due to the low TOC contents samples from below 19 m bls, nearly no information is available for facies F3a and F2.

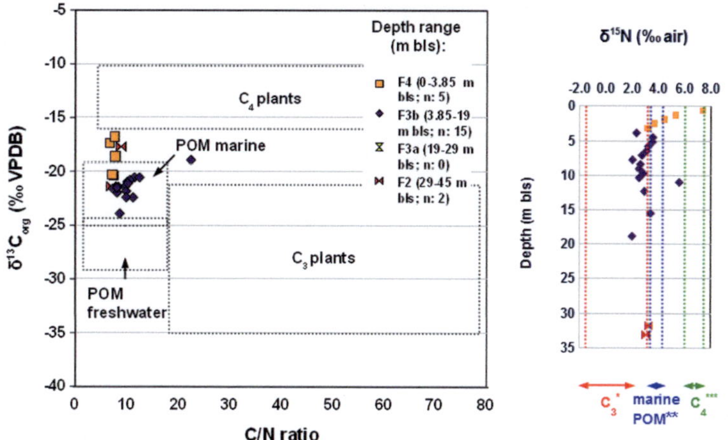

Figure 6.6: Distribution of C/N ratios and δ^{13}C and δ^{15}N values in comparison to characteristic ranges for different OM sources (n: 22).

*δ^{15}N range for mangrove (C$_3$ type) tissues (MUZUKA & SHUNULA 2006)

**δ^{15}N range for sedimentary POM from the Bay of Bengal (GAYE-HAAKE et al. 2005)

***δ^{15}N range for C$_4$ plant pollen (DESCOLAS-GROS & SCHÖLZEL 2007)

6.2.2 GROUNDWATER CHARACTERISATION

6.2.2.1 Groundwater properties

Monitoring wells. Physico-chemical properties determined from recorded data of the five monitoring wells before the in-situ experiment was performed in December 2009 are summarised in Table 6.3. These results are considered to reflect representative baseline values for the low As site two months after the monsoon season ended. To identify potential outliers, results were compared to previously taken monitoring samples.

Table 6.3, part I: Selection of hydrochemical baseline values that are used in the following to interpret changes in the groundwater composition (complete results presented in APPENDIX II, Table A 2.5). Samples were taken directly before the sucrose experiment was conducted (03/12/09). Table continued on next page.

Well	pH	EC	Temp.	TA HCO_3^-	Na Na^+	K K^+	Ca Ca^{2+}	Mg Mg^{2+}
(m bls)		(µS/cm)	(°C)	(mg/L)	(mg/L)	(mg/L)	(mg/L)	(mg/L)
A (12-21)	7.1	647	27.0	415	18.6	2.49	86.5	20.6
B (24-27)	7.2	623	27.0	482	12.5	3.79	79.3	17.3
C (30-33)	7.2	624	26.8	451	11.4	6.82	69.5	17.4
D (36-39)	7.2	604	26.5	464	10.8	3.72	67.6	15.8
E (42-45)	7.1	590	26.8	445	11.1	3.18	68.0	15.1
Aqueous/solid phase ratio range				nd	1.09 to 1.86 $\times 10^{-3}$	1.73 to 4.80 $\times 10^{-4}$	3.43 to 6.23 $\times 10^{-3}$	3.32 to 5.94 $\times 10^{-3}$

Table 6.3, part II: Negative ion balances are attributed to inaccuracies of the titrimetric alkalinity field kit. Thermodynamically dominating species according to PHREEQC calculations.

Well (m bls)	Cl^- (mg/L)	SiO_2 (mg/L)	DOC (mg/L)	$\delta^{18}O$ (‰ VSMOW)	δ^2H	O_2 (mg/L)	NO_3^- (mg/L)	NH_4^+ (mg/L)	Ion balance (%)
A (12-21)	7.85	34.8	7.47	-4.35	-29.2	error	<0.88	1.68	+0.90
B (24-27)	2.65	31.5	6.68	-4.45	-28.7	0.53	<0.88	3.01	-9.90
C (30-33)	2.43	28.5	5.10	-3.99	-28.3	0.33	<0.88	3.67	-11.4
D (36-39)	2.64	26.6	4.73	-4.28	-29.4	0.24	<0.88	3.39	-14.2
E (42-45)	2.59	25.5	6.67	-4.55	-30.6	0.18	<0.88	3.03	-14.8
Aqueous/ solid phase ratio range	nd	nd	nd			nd	nd	nd	

Well (m bls)	Mn Mn^{2+} (mg/L)	Fe Fe^{2+} (mg/L)	SO_4^{2-} (mg/L)	PO_4^{3-} $H_2PO_4^-$ (mg/L)	As* H_3AsO_3 (µg/L)	Zn Zn^{2+} (µg/L)	Co Co^{2+} (µg/L)	Ni $NiCO_3$ (µg/L)	V $H_2VO_4^-$ (µg/L)
A (12-21)	0.77	4.34	<0.85	2.38	49.1	*7.48*	4.33	*0.52*	*0.08*
B (24-27)	0.47	3.76	<0.85	3.39	155	3.05	0.41	0.51	0.12
C (30-33)	0.41	5.57	<0.85	1.12	135	4.50	0.29	0.36	0.02
D (36-39)	0.42	2.86	<0.85	2.18	132	5.79	0.37	0.37	0.08
E (42-45)	0.60	2.15	<0.85	1.93	133	5.35	0.41	0.37	0.06
Aqueous/ solid phase ratio range	0.88 to 1.92 $\times 10^{-3}$	1.46 to 5.34 $\times 10^{-3}$	nd	1.51 to 3.18 $\times 10^{-3}$	2.03 to 9.12 $\times 10^{-2}$	1.22 to 2.04 $\times 10^{-4}$	0.60 to 9.45 $\times 10^{-4}$	3.55 to 6.37 $\times 10^{-5}$	0.96 to 4.11 $\times 10^{-6}$

*Average As(III) percentage: 95.8 %

nd: not determined

Outlier, replaced by monitoring result from 25/11/09

Results reveal that noticeable concentrations of Fe, Mn, PO_4^{3-}, and As are in solution, while O_2, NO_3^- and SO_4^{2-} are virtually absent. Aqueous to solid phase ratios for different elements were determined from groundwater and sediment samples of corresponding depth intervals (Table 6.3 and APPENDIX II, Table A 2.4) and reflect a striking enrichment of As in groundwater. On 03/12/09, the hydrostatic head was met at 1.81 m bls in the monitoring wells, which is clearly above the base of the silty and clayey layers (reaching down to 3.20 m bls). Hence, the aquifer was in a confined condition to this time. Results reveal that the anoxic groundwater is characterised by circum-neutral pH values, a high alkalinity and belongs to the Ca-(Mg)-HCO_3-type (Figure 6.7).

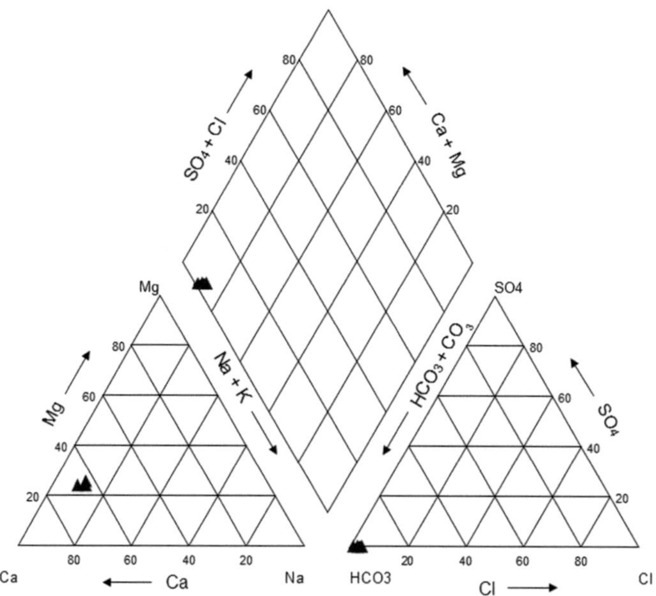

Figure 6.7: Piper diagram presenting the major solute composition in groundwater of the five monitoring wells (03/12/09).

Values of $\delta^{18}O$ and δ^2H appear as comparatively indifferent in contrast to the high As site (see Figure 7.8, chapter 7.2.2.1). Values plot consistently to the right of the local meteoric waterline (LMWL), which is according to GAT (1996) a strong evidence for previous evaporation influences (Figure 6.8). According to hydrochemical conditions, SI further display that groundwater is supersaturated regarding magnetite, hematite, siderite, calcite and ordered dolomite, whereas disordered dolomite is undersaturated (see Table 6.4).

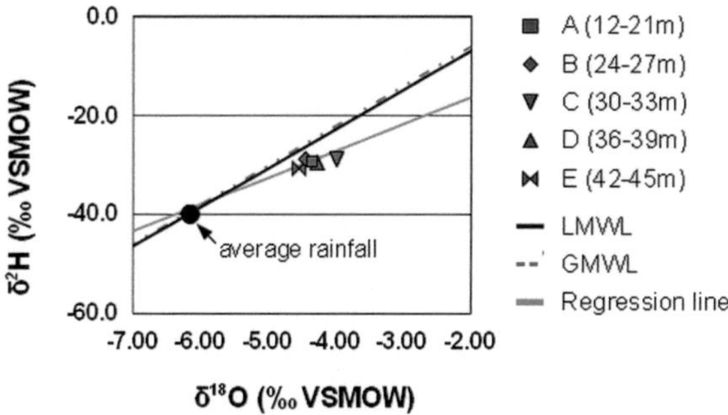

Figure 6.8: Isotopic compositions of O and H in samples (03/12/09) compared to the global meteoric water line (GMWL), the local meteoric water line (LMWL), and the volume-weighted average composition of rainfall in 2004/05 (in SENGUPTA et al. 2008) for an area situated about 40 km south of the study site (JAM study site).

Table 6.4: Saturation indices for detected and potentially present mineral phases, calculated on the basis of field measurements and analyses of samples taken on the 03/12/09 (redox assumptions: O_2: 0.24 mg L^{-1}, E_H: -85 mV).

Well (m bls)	Quartz	Anorthite	Calcite	Dolomite ord./ disord.		Chlorite	Phlogopite
A (12-21)	0.38	-1.96	0.20	0.15	-0.39	-7.85	-8.52
B (24-27)	0.34	-2.67	0.29	0.30	-0.24	-8.39	-8.57
C (30-33)	0.30	-5.68	0.19	0.16	-0.38	-11.6	-9.99
D (36-39)	0.28	-4.47	0.19	0.12	-0.42	-10.6	-9.82
E (42-45)	0.25	-4.73	0.16	0.03	-0.51	-11.2	-10.3

Well (m bls)	Fe(OH)$_3$ amorph.	Magnetite	Hematite	Siderite ord./ disord.		Halite
A (12-21)	-2.45	11.1	9.04	1.00	0.55	-8.42
B (24-27)	-2.33	11.4	9.28	1.05	0.60	-9.07
C (30-33)	-2.21	11.8	9.50	1.19	0.74	-9.14
D (36-39)	-2.51	10.9	8.88	0.91	0.46	-9.13
E (42-45)	-2.71	10.3	8.51	0.75	0.30	-9.12

Recorded abundances of microbes as indicated by the presence of free and cultivatable germs in water range between 3.4 to 6.3 × 10^5 cfu mL^{-1} (24 h incubation at 30°C), with no evidence of pathogenic germs in general, and for *Escherichia coli* in particular. In agreement with the computed E_H-pH-diagram for the local groundwater composition (Figure 6.9), As(III) is with a percentage of >95.8 % the predominating As species.

The hydrochemical composition of groundwater from the shallowest well A differs significantly from the deeper wells. Concentration differences comprise Cl$^-$, Na, Ca, Co, NH$_4^+$, and importantly, As and As(III). Hence, results point at the presence of two distinctive hydrological layers within the groundwater body, which are divided by a boundary located in between the well screens of wells A and B (in about 21 to 24 m bls).

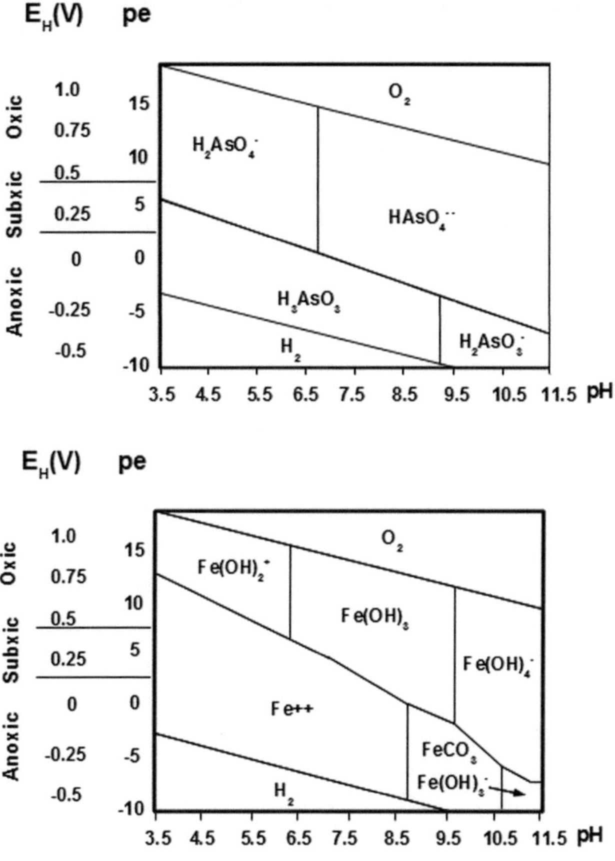

Figure 6.9: E_H-pH diagram reflecting the As species distribution within the As-O_2-H_2O-system and the Fe species distribution in the Fe-O_2-H_2O-CO_3 system in solution, based on the groundwater composition in well B (03/12/09). At the study site, H_3AsO_3 and Fe(II) are the dominating species (conditions: 27°C, 0.1 MPa, pH ranging from 3.5 to 11.5, pe ranging from +17.5 to -10, created with PHREEQC; based on MERKEL & PLANER-FRIEDRICH 2008).

Monitoring wells and adjacent tube wells. In the following, the five monitoring wells are compared to adjacent local tube wells in order to estimate their representativity for the study area. No significant differences are visible between principal groundwater properties in the monitoring wells of the low As site and 11 adjacent tube wells situated in the villages Chakudanga and Amdanga (Table 6.5). Depending on the concentrations of redox sensitive parameters in groundwater (NO_3^-, Mn, Fe, and SO_4^{2-}), six wells are considered to be in a less reducing state of $Fe(III)/SO_4^{2-}$ reduction than the monitoring wells. Two wells (wells 118 and 119) hold extremely low As and PO_4^{3-} concentrations. According to the high similarity with the majority of the surrounding wells, groundwater samples from the low As study site are considered as representative for the local area.

Results for well 125, which was sampled during the field survey (conducted between September to November 2007) are of special interest, since this well was the reason to choose this area as low As study site. Although nearly all major and trace elements are comparable to the monitoring wells, NO_3^-, PO_4^{3-}, Fe, and As significantly differ from groundwater of the monitoring wells (well 125a in Table 6.5). Hence, the well was sampled again in December 2009, revealing that results were now comparable to those of the deeper monitoring wells, except for the higher Cl^- and SO_4^{2-} concentrations (well 125b in Table 6.5). The reason for this temporal change is very likely an on-going activity of microbially mediated redox reactions (discussed in chapter 6.3.4).

Table 6.5: Comparison of monitoring wells (03/12/09) with nearby wells that were sampled during the field survey in 2007. Well 125 formed the base of decision to choose this area as low As study site. This well was sampled again in 2009 (125b).

Well (m bls)	Year	Ca (mg/L)	Mg (mg/L)	Na (mg/L)	K (mg/L)	Cl^- (mg/L)	NO_3^- (mg/L)	Mn (mg/L)	Fe (mg/L)	SO_4^{2-} (mg/L)	PO_4^{3-} (mg/L)	As (µg/L)	As/PO_4^{3-} (mol ratio)	Na/Cl^- (mol ratio)
						Monitoring wells:								
A (12-21)	2009	86.5	20.6	18.6	2.49	7.85	<0.88	0.77	4.34	<0.85	2.38	49.1	0.03	2.36
B (24-27)	2009	79.3	17.3	12.5	3.79	2.65	<0.88	0.47	3.76	<0.85	3.39	155	0.06	4.72
C (30-33)	2009	69.5	17.4	11.4	6.82	2.43	<0.88	0.41	5.57	<0.85	1.12	135	0.15	4.70
D (36-39)	2009	67.6	15.8	10.8	3.72	2.64	<0.88	0.42	2.86	<0.85	2.18	132	0.08	4.11
E (42-45)	2009	68.0	15.1	11.1	3.18	2.59	<0.88	0.60	2.15	<0.85	1.93	133	0.09	4.31
						Adjacent wells:								
125a (24)	2007	109	24.1	15.1	4.03	46.7	1.68	0.51	0.05	10.7	0.03	2.29	0.10	0.50
125b (24)	2009	82.5	20.3	12.3	3.97	56.6	<0.88	0.41	4.18	12.4	5.95	136	0.03	0.33
124 (24)	2007	85.9	16.4	13.8	3.69	3.55	<0.88	0.47	4.85	<0.85	2.58	160	0.08	5.99
126 (24)	2007	71.3	14.3	10.8	3.01	3.56	1.38	0.27	1.04	5.42	3.83	61.1	0.02	4.66
122 (167)	2007	71.0	14.0	9.46	2.36	2.25	<0.88	0.06	1.07	<0.85	0.41	77.4	0.24	6.49
Well east (24)	2009	69.9	15.1	11.7	3.51	11.7	<0.88	0.35	3.54	<0.85	5.26	124	0.03	1.55
Pumping well (24)	2009	72.4	15.5	11.0	2.36	8.09	<0.88	0.28	4.84	<0.85	5.64	63.7	0.01	2.10
117 (30)	2007	83.5	20.9	15.6	2.20	21.9	1.72	0.37	6.54	16.27	4.16	37.9	0.01	1.10
118 (?)	2007	105	20.2	9.65	3.71	14.3	1.17	0.57	13.9	3.91	0.57	4.62	0.01	1.04
119 (18)	2007	113	16.4	9.29	2.65	10.7	1.43	0.40	1.38	4.51	0.14	3.24	0.03	1.34
121 (25)	2007	80.8	20.9	10.8	2.72	10.9	<0.88	0.18	4.05	2.00	3.79	50.4	0.02	1.52
120 (183)	2007	108	21.2	14.3	3.42	4.96	1.88	0.06	1.30	<0.85	0.07	34.9	0.63	4.46

6.2.2.2 The sucrose injection experiment

Preliminary experiments. The fundamental design of the in-situ sucrose injection experiment bases on a field experiment done by HARVEY et al. (2002) and laboratory incubation experiments that have examined the role of geomicrobiological related processes in As mobilisation (DEUEL & SWOBODA 1972). Anaerobic microbial column experiments conducted by D. FREIKOWSKI with sediment material from both sites demonstrated that indigenous microbes are still capable of As mobilisation as soon as OM is provided (FREIKOWSKI et al., unpublished results).

Consequences of circular pumping. In order to equally distribute the dissolved sucrose that was added into the four surrounding wells (B to E), groundwater was extracted from the central well A and directly re-introduced into the surrounding wells one after another (see Figure 4.3 in chapter 4.3.2). This circular pumping successfully distributed the sucrose within the local aquifer, as demonstrated by sucrose concentrations determined two days after infusion (Table 6.6). However, nearly no sucrose occurred in groundwater from wells A and C. Since no sucrose was introduced into well A, concentrations remained throughout low (Figure 6.10). In contrast, the initially low concentration in well C remains obscure, since the same amount of sucrose was inserted than into well D. The next sample taken four days later contained considerable 486 mg L^{-1} of sucrose (APPENDIX II, Table A 2.5).

Samples taken two days after sucrose addition (06/12/09) displayed influences on the local groundwater composition related to circular pumping, which caused mixing of the hydrochemically layered groundwater. In the central well A, As concentrations temporarily increased from initially 46.2 to 66.4 µg L^{-1}, while major solutes declined (e.g., Cl$^-$ and Na). Contrary effects occurred in groundwater of the deeper monitoring wells as indicated by concentration changes, for example, in conservative Cl$^-$. By contrast, PO$_4^{3-}$ contents had multiplied in wells C and E, and with a lag of two days in the other three wells, too.

A significant input of As and Fe by sucrose was excluded, 500 mg L^{-1} solution contained 0.94 µg L^{-1} As and 3.70 µg L^{-1} Fe, only.

Table 6.6: Sucrose concentrations in groundwater from the monitoring wells two days after sucrose addition reflect a successful distribution within the aquifer, except for well A (where no sucrose was added) and well C.

Well (well screen m bls)	Date	Sucrose (mg/L)	(mM)
A (12-21)	06/12/09	8.20	0.02
B (24-27)	06/12/09	290	0.85
C (30-33)	06/12/09	6.23	0.02
D (36-39)	06/12/09	259	0.76
E (42-45)	06/12/09	873	2.55

Temporal changes in groundwater following sucrose injection. In addition to effects of circular pumping, the hydrochemistry in groundwater of the study site changed rapidly in the following days, with an intensity and duration of response directly depending on the available sucrose contents (Figure 6.10).

The following changes manifested in the groundwater composition:

- Well A: Around the central well A, the throughout low sucrose concentrations provoked comparatively weak changes in the groundwater composition, except for the effects related to circular pumping during sucrose addition.

- In the four other wells, most parameters developed trends peaking 6-8 days after injection, while sucrose concentrations gradually decreased and completely vanished after 14 days.

- Well E: Strongest reactions developed in well E, which showed the highest initial sucrose concentration. Here, temporary drastic increases in conductivity and alkalinity arose, while fatty acids (primarily acetate, to a lesser extent butyrate and propionate, see APPENDIX II, Table A 2.5) occurred. This was accompanied by a temporary drop in the pH by two units down to 5.0, changing the groundwater chemistry from circum-neutral to weakly acidic. With a lag of two days, Fe concentrations rose to a maximum (77.8 mg L^{-1}) that was 36-times higher than the baseline (2.15 mg L^{-1}). Dissolved Mn concomitantly increased from 0.59 to 4.47 mg L^{-1} (x 7.48; corresponds to a maximum multiplication factor of 7.48), while concentrations of trace elements witnessed partly also steep raises (Zn x 78.0, Co x 47.3, Ni x 36.5, and V x 33.1). Furthermore, sharp increases in alkaline earth metals (Ca x 3.83, Mg x 2.42, Ba x 5.20, and Sr x 3.37) and a moderate release of Si (x 1.47) appeared, forming similar trends to those of Fe. Conductivity nearly tripled (from 590 to 1,700 µS cm^{-1}), while alkalinity doubled (from 445 to a maximum of 976 mg L^{-1}).

- Wells B, C and D: Hydro-chemical parameters evolved similar in wells B and C, where highest sucrose concentrations reached 33.2 % (B), respectively 53.6 % (C), of that in well E. The gradual decline of sucrose in groundwater was accompanied by similar temporary effects previously described for well E. Fatty acids emerged, pH values declined and concentrations of Ca, Fe, Mn, and other trace elements rose. In contrast, trends in well D (initial sucrose concentration was 29.7 % of that in well E) differed from those observed in the other wells. Here, characteristic parameters like dissolved Ca and Fe increased slowly and peaked very late.

- All wells: Dissolved O_2 in groundwater permanently remained below 0.5 mg L^{-1}, and neither NO_3^- nor SO_4^{2-} was detectable. Reduced Fe(II) was consistently the prevailing Fe species, with an average percentage of 96.7 ± 3.3 % (n: 30). The presence of free germs escalated by several orders of magnitude and demonstrated a dramatic growth of the microbial population within the aquifer. Four days after sucrose infusion, the appearance of foam and an acidic smell of fermentation was noticed during sampling.

Hydrochemical parameters that developed trends in wells B, C and E peaked between six (10/12/09) to eight days (12/10/12) after sucrose infusion. To this time, the approximate redox potentials were measured, allowing calculations of respective SI for Fe-minerals (Table 6.7). In addition, partly high concentrations of dissolved CO_2 were calculated from TA. After peaking, most parameters developed rapid trends in direction of the initial baseline values. Nevertheless, a few strongly increased parameters (e.g., TPC, Fe, TA, Co and PO_4^{3-}) stagnated at first, or declined only very slowly.

Arsenic. Arsenic concentration trends matched with those of other elements and parameters, but the relative increases were throughout lower. In groundwater of all wells, ratios of As to PO_4^{3-} as well as As to Fe continuously declined. Maximum relative increments of dissolved As reached 19.0 % (well C), 23.0 % (well D), 39.9 % (well B) and 48.6 % (well E) in respect to the initial baseline values. As(III) remained with an average percentage of 95.5 ± 2.8 % (n: 37) the predominating As species. Except for well B, total As and As(III) concentrations rapidly returned to the respective baseline concentrations after maximum concentrations were reached.

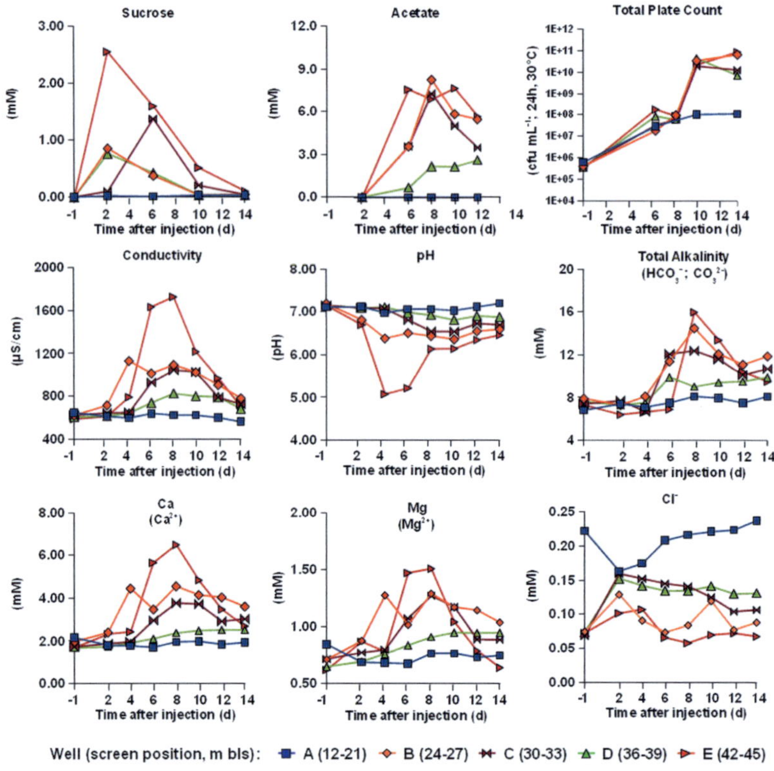

Figure 6.10, part I: Effects on the local hydrochemistry during the following 14 days after biostimulation with sucrose. Included are thermodynamically favoured species for each solute. Figure continued on next page.

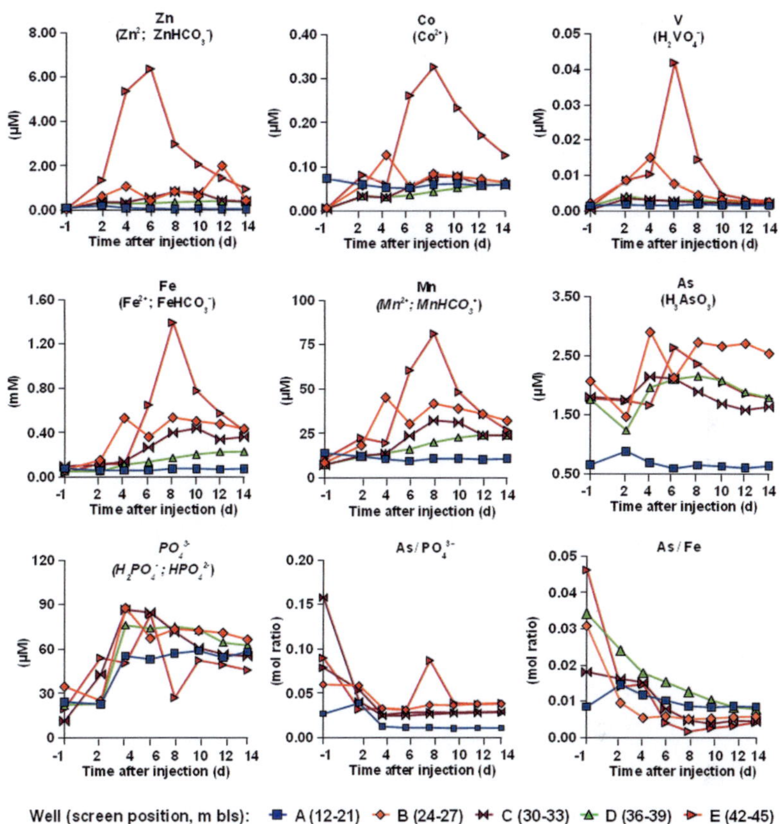

Figure 6.10, part II: Effects on the local hydrochemistry during the following 14 days after biostimulation with sucrose. Included are thermodynamically favoured species for each solute. For a better comparability of the trends, concentrations are given in molar concentrations. Strongest temporary mobilisation reactions occurred in the deepest well E, where highest initial sucrose concentrations prevailed.

Table 6.7: Results and computed SI of detected and potentially relevant mineral phases. Calculations based on samples taken six days after sucrose addition (10/12/09), except for well E, here eight days later (12/12/09). E_H measurements are mandatory to calculate SI for redox dependent mineral phases. Computed HCO_3^- and CO_2 distributions were calculated from measured alkalinity, pH and temperature with PHREEQC.

Well (m bls)	pH	Eh (mV)	EC (µS/cm)	O_2 (mg/L)	NO_3^- (mg/L)	SO_4^{2-} (mg/L)	PO_4^{3-} (mg/L)	Fe (mg/L)	Mn (mg/L)	As (µg/L)
A (12-21)	7.1	-85.9	638	error	<0.88	<0.85	5.22	3.25	0.52	44.5
B (24-27)	6.5	-290	1013	0.25	<0.88	<0.85	6.61	20.2	1.67	159
C (30-33)	6.8	-291	928	0.14	<0.88	<0.85	8.31	14.9	1.30	158
D (36-39)	7.0	-190	738	0.11	<0.88	<0.85	7.24	7.70	0.89	158
E (42-45)	6.2	-384	1705	error	<0.88	<0.85	2.65	77.8	4.47	177

Well (m bls)	HCO_3^- measured (mg/L)	HCO_3^- calculated (mg/L)	CO_2 calculated (mg/L)	Sucrose (mg/L)	Acetate (mg/L)	Butyrate (mg/L)	Propionate (mg/L)
A (12-21)	458	443	54.2	3.30	<dl*	<dl	<dl
B (24-27)	695	661	500	126	210	22.2	33.7
C (30-33)	738	705	151	468	210	<dl	<dl
D (36-39)	604	582	79.5	145	39.0	<dl	<dl
E (42-45)	976	888	817	nd	406	34.8	124

Well (m bls)	Quartz (SI)	Anorthite (SI)	Calcite (SI)	Dolomite disord. (SI)	$Fe(OH)_3$ amorph. (SI)	Magnetite (SI)	Hematite (SI)	Goethite (SI)	Siderite disord. (SI)	Vivianite (SI)
A (12-21)	0.29	-4.06	0.11	-0.57	-2.71	10.4	8.53	3.26	0.42	1.91
B (24-27)	0.88	-2.53	0.02	-0.96	-11.8	-7.59	-8.83	-5.55	0.75	0.40
C (30-33)	0.36	-4.85	0.23	-0.36	-6.37	3.10	1.17	-0.42	0.93	3.12
D (36-39)	0.30	-4.97	0.23	-0.32	-4.31	7.37	5.29	1.63	0.80	2.93
E (42-45)	0.41	-4.60	0.03	-1.07	-9.22	-3.16	-4.48	-3.25	1.03	1.74

6.2.2.3 Monitoring results

The bi-weekly taken monitoring samples cover a period of 20 months (January 2009 to August 2010) and comprise 35 times of sampling (Figure 6.11). This elaborate data set is used in the present thesis to trace ground-water evolution notably in respect of As mobility, and to determine mid-term influences of the sucrose experiment. Records of the hydrostatic head revealed pronounced oscillations during this period, which were closely linked to prevailing climatic conditions. Hydrostatic head positions in well A ranged from 0.39 m bls (end of the monsoon season 2009), to 4.35 m bls (end of the pre-monsoon season in April 2010).

Time resolved variations in the groundwater hydrochemistry during 2009. Changes in the hydrostatic head were accompanied by pronounced variations in different parameters recorded in the shallowest well A (Figure 6.11). During the dry pre-monsoon season, elements with deviant concentrations from the deeper wells like As, As(III) and PO_4^{3-} increased, respectively decreased considerably at the same time (e.g., Na, Cl$^-$). During the following monsoon season, opposite trends appeared and respective concentrations stabilised rapidly at new levels until end of May.

Until sucrose was introduced in December 2009, most major and trace element concentrations in the four other monitoring wells remained stable, except for $\delta^{18}O$ values, PO_4^{3-} concentrations and As(III) percentages. These specific parameters decreased during the pre-monsoon season to increase again with onset of the monsoon rains. In all samples, NO_3^- concentrations remained constantly below the limit of quantification, except for some sporadic outliers observed at the beginning of the monitoring.

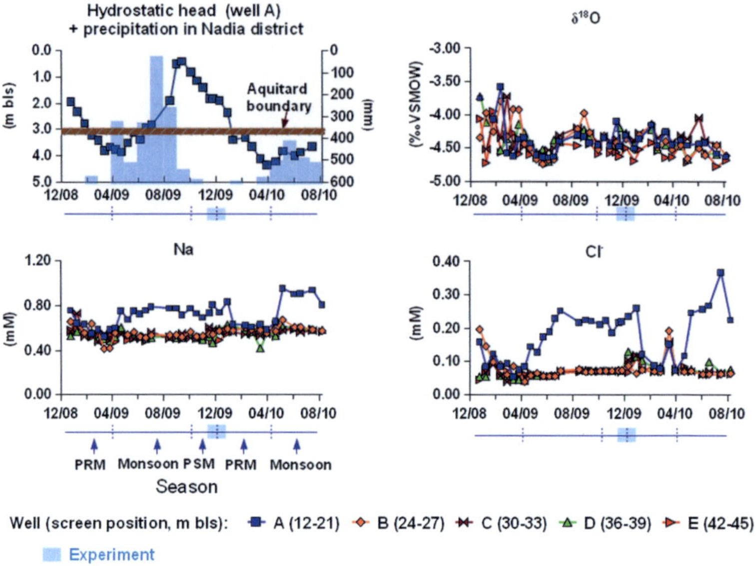

Figure 6.11, part I: Summary of hydrochemical parameters relevant for As mobility from monitoring samples taken between 06/01/09 and 12/08/10 (PRM: pre-monsoon season; PSM: post-monsoon season). Average monthly precipitation values for the Nadia district from 2008 to 2010 provided by the Hydromet Division of the India Meteorological Department. Figure continued on next page.

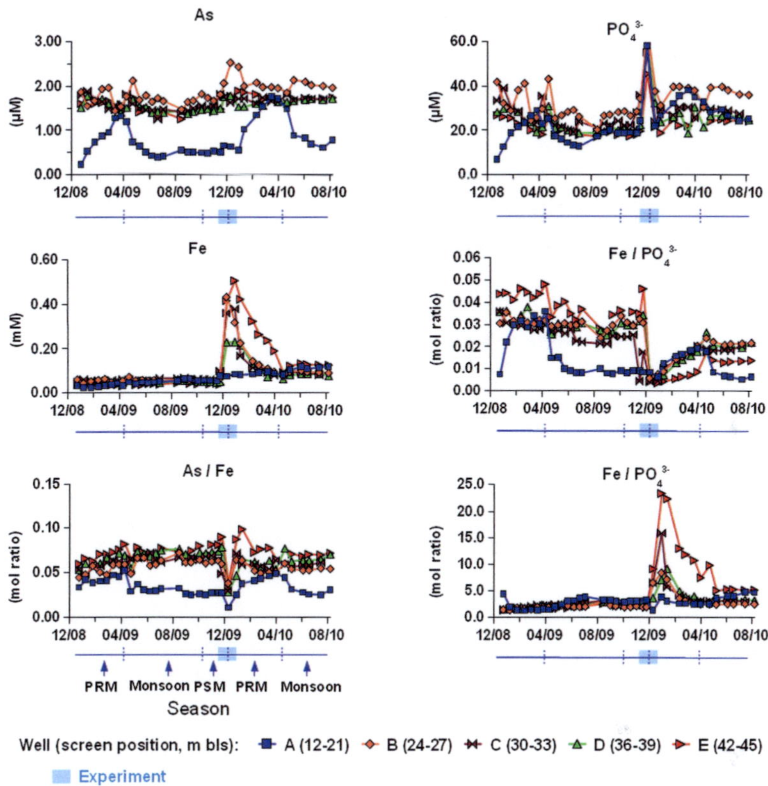

Figure 6.11, part II.

Mid-term effects following the sucrose experiment. In January, characteristic parameters like Fe and As finally peaked in well D, too. Generally, parameters developed steady decreases towards the baseline values (e.g., Fe, Zn, Ni, and Co) (Figure 6.12).

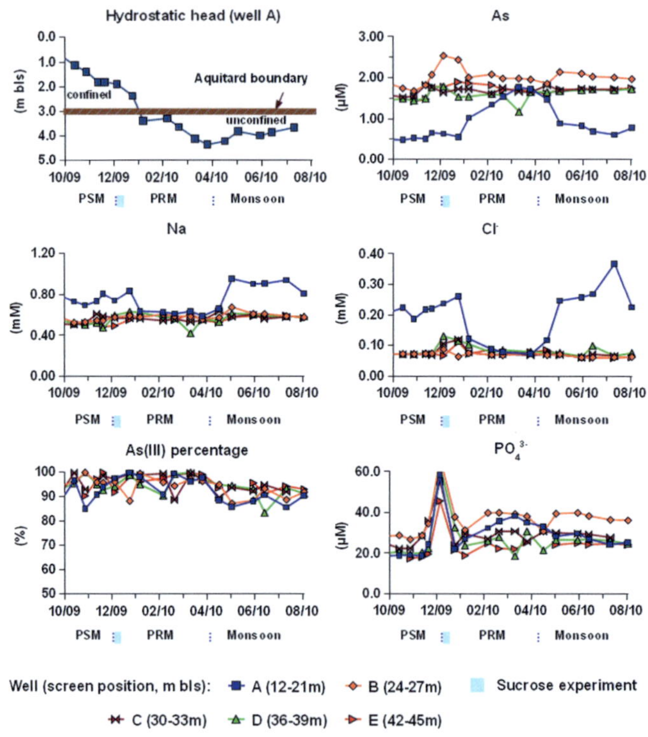

Figure 6.12, part I: Summary of hydrochemical trends relevant for As mobility following the biostimulation experiment. Parameters that increased by sucrose addition continuously declined until beginning of the monsoon season. Additionally, characteristic elements (e.g., As, Na, Cl⁻, and PO₄³⁻) developed trends in well A that were closely linked to oscillations of the hydrostatic head (PSM: post-monsoon season; PRM: pre-monsoon season). Figure continued on next page.

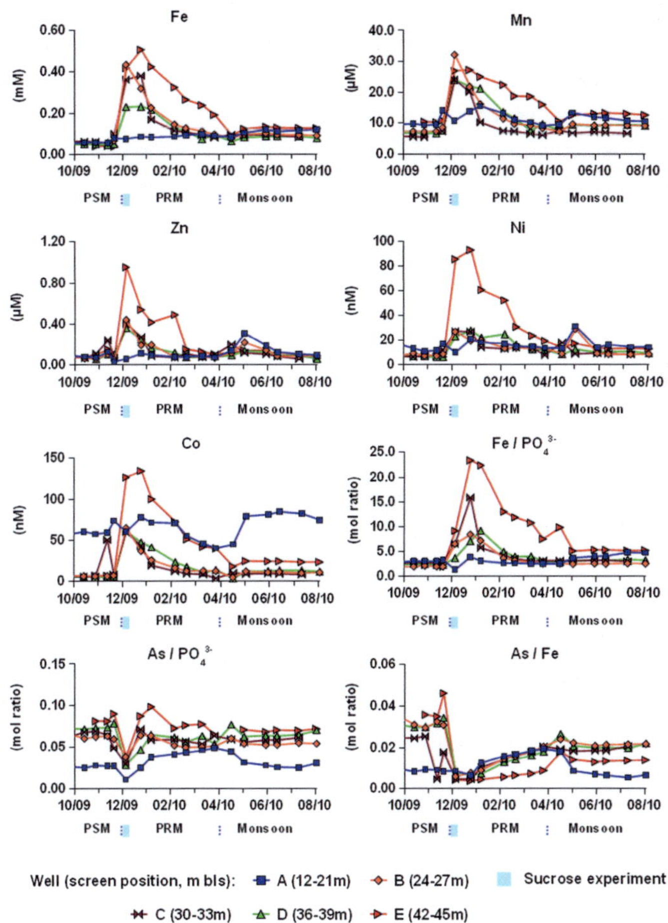

Figure 6.12, part II: Parameters that increased after addition of sucrose continuously declined until beginning of the monsoon season (PSM: post-monsoon season; PRM: pre-monsoon season).

At the beginning of the monsoon season end of April 2010, the hydrostatic head slowly increased. Again, pronounced trends in PO_4^{3-}, As, Na, Cl^-, and Co concentrations occurred in well A, similar to the values at the beginning of the monsoon season in 2009. This time, some parameters changed in the other wells, too, especially Fe increased again and levelled off at concentrations above the initial baseline. Additionally, As abruptly increased in well B, before declining again. Compared to average concentrations recorded shortly before sucrose was infused, As and Fe concentrations had increased in all of the wells (Figure 6.13).

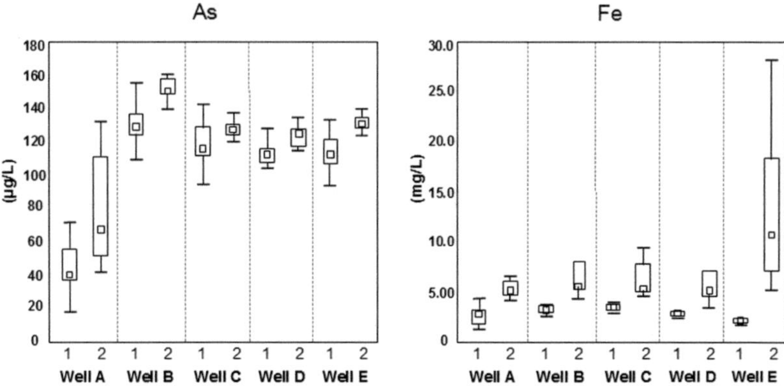

Figure 6.13: Summary of As and Fe concentrations before (1: 06/01/09-03/12/09; n: 22) and after (2: 18/12/09 – 12/08/10; n: 13) the experiment in form of Box-Whisker-Plots.

Constant arsenic increase in well A. The short termed, seasonal increases and decreases of As concentrations in well A were superimposed by a constantly increasing trend (Figure 6.14). Iron and PO_4^{3-} concentrations paralleled this increase, while conservative tracers Cl^- and $\delta^{18}O$ and most other parameters did not follow this trend. In order to quantify the net increases in As, PO_4^{3-} and Fe concentrations, the seasonal minima in July 2009 and 2010 are compared to the first sample taken in January 2009 (Table 6.8). In contrast to changes in absolute As concentrations, As(III) remained constantly the prevailing As species. Groundwater from the central well A further differed from the deeper wells in respect of the mol ratios of As, Fe and PO_4^{3-} (Figure 6.15).

Table 6.8: Concomitant net increases of As, Fe and PO_4^{3-} in groundwater of well A to the begin of the monitoring and respective As minima in July 2009 and July 2010.

Date	Mn (mg/L)	Fe (mg/L)	PO_4^{3-} (mg/L)	As (µg/L)	As to PO_4^{3-} (mol ratio)
06/01/09	0.60	1.73 (baseline)	0.68 (baseline)	17.6 (baseline)	1 : 30
14/07/09	0.67	2.78 (+61 %)	1.25 (+84 %)	30.5 (+73 %)	1 : 31
21/07/10	0.59	6.44 (+272 %)	2.37 (+249 %)	45.8 (+160 %)	1 : 40

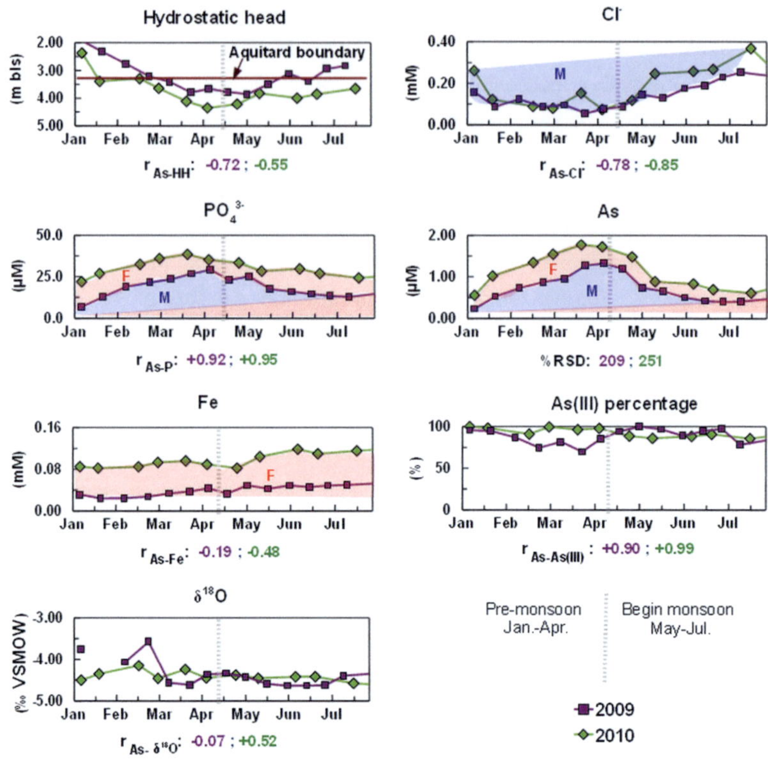

Figure 6.14, part I: Comparison between monitoring results for well A from 2009 and 2010. In order to distinguish between permanent net increases and temporary, seasonal concentration changes, areas below the trend lines are shown in red (F: net increase) and blue (M: seasonal effect). Net increases refer to the first sample taken in January 2009. Figure continued on next page.

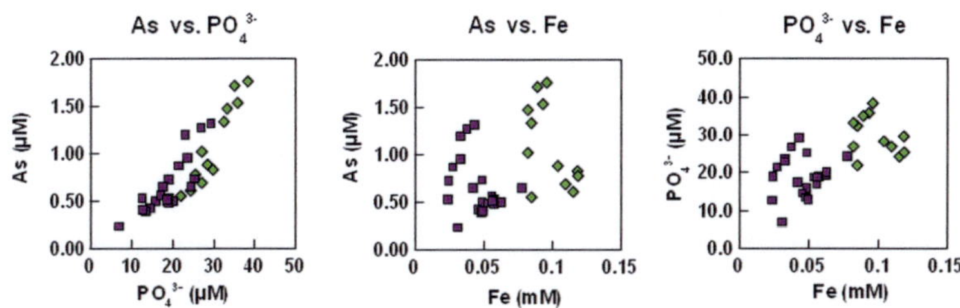

Figure 6.14, part II: Mol ratios of As, PO$_4^{3-}$ and Fe for monitoring samples from well A (2009: purple, 2010: green).

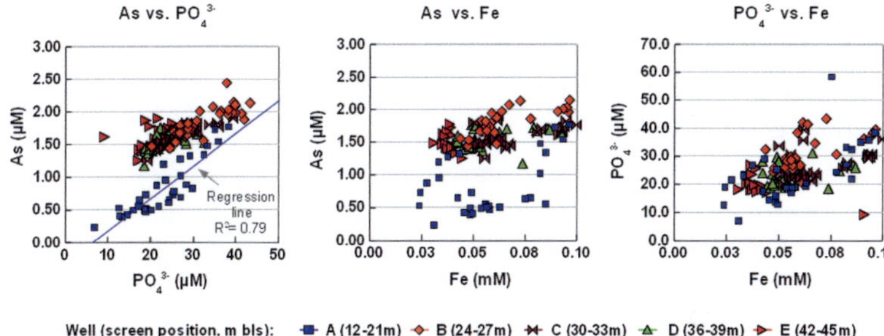

Figure 6.15: As, Fe and PO$_4^{3-}$ in monitoring samples (06/01/09 to 12/08/10) display strong positive correlations and contrasting concentrations in well A. The partial decoupling of As from Fe in well A originates from the seasonal fluctuations of the As concentrations.

6.3   DISCUSSION

6.3.1   SEDIMENT STRATIGRAPHY

Following considerations regarding the aquifer characteristics at the low As site are based on data from well E, where the sediment samples were collected. Since the distances to the other four wells are less than 4-5 m, similar conditions can be assumed for the whole study site.

When investigating the fate of a highly versatile element such as As, it is mandatory to understand the investigation area's evolution history and geomorphologic background. Local sediments witness various significant changes during the evolution of the floodplain. Present lithology was and still is strongly influenced by the nearby Hooghly River, which is currently located 12.6 km west of the low As site. Additionally, regression and transgression events that occurred during the late Quaternary had decisive influences on the regional sedimentation history (see chapters 3.1 and 3.2). By combining geochemical and lithological data with characteristics of the sedimentary OM, further conclusions can be drawn concerning the delta and floodplain evolution. Together with results from the high As site, the facies can be interpreted in context of a fluvial sequence model (GIBLING & BIRD 1994, MIALL 1996, SHANLEY & MC CABE 1994, WRIGHT & MARRIOTT 1993). The outcomes are further compared to the nearby JAM study site, where a similar approach was persuaded and included additionally radio-carbon and optically stimulated luminescence (OSL) dating (MC ARTHUR et al. 2008, SARKA et al. 2009).

Interpretations and outcomes summarised in Table A 2.6 (APPENDIX II) are in good agreement with those of the JAM site, except for contrary conclusions regarding past marine influences. SARKA et al. (2009) excluded marine influences on the bases of strong variations in $\delta^{15}N$ signals (-3.54 to +2.76 ‰) of a set of six analysed samples only. Although both study sites are located approximately 40 km north of the JAM area, OM in facies F2 and F3 carries here marine signals as visible in $\delta^{15}N$ and $\delta^{13}C$ signals as well as C/N ratios (Figure 6.6). According to LAMB et al. (2006), the mixing of marine and terrestric signals is in general characteristic for estuarine systems like salt marshes or lagoons. Hence, sediment samples are considered to indicate a subsequent transition from a marine-deltaic environment towards a more terrestric influenced floodplain environment during the early to middle Holocene (see APPENDIX II, Table 2.8). Comparing the outcomes of this sequence stratigraphic analysis to the dating results of SARKA et al (2009), the conclusion is drawn that increased As concentrations in groundwater of both sites are closely linked to Holocene and late Pleistocene aquifer sediments.

6.3.2 ROLE OF THE SURFACE AQUITARD

The unconsolidated aquifer sediments are covered by the 3.85 m thick facies F4, which is interpreted as fluviatile overbank deposit of the Hooghly River (see chapter 6.2.1). These sediments are considered to act rather as an aquitard than an aquiclude and are characterised by low hydraulic conductivities. This unit is a common feature of the floodplain, where it occurs in varying thicknesses of up to ~24 m and is locally flooded during the monsoon season, similar to some areas in Bangladesh (MC ARTHUR et al. 2004).

Between the riverbanks of the Hooghly River and Chakdah, mildly to strongly reducing conditions were reported from the shallow aquifer parts beneath this clay-rich layer, which reaches thicknesses of 5 to 15 m here (CHARLET et al. 2007). Within the aquitard, small-scaled sandy lenses with diameters of several decimetres were detected with electromagnetic conductivity measurements (MÉTRAL et al. 2008). These lenses in the aquitard enable locally restricted infiltration of oxygen- and nitrate-rich surface water, which prevents a decrease of the redox potential in the underlying aquifer (MÉTRAL et al. 2008). This is a very important observation, giving some clues to understand how the aquifer is recharged during the monsoon season, and why the hydrochemistry is often spatially varying within distances of decimetres.

At the low As site, the clay layer obviously prevents the inflow of oxygen-rich surface water, as indicated by groundwater chemistry that indicates moderate to strongly reducing conditions in the aquifer. This also hinders the import of DOC and PO_4^{3-} from adjacent fields. In addition, clinker fragments and compacted sediments in the top 0.60 m reflect an anthropogenic influence close to the surface, while sediments below are considered as undisturbed.

Seasonal changes in the hydrostatic head induce switches between unconfined and confined conditions as soon as the water table reaches the clayey and silty aquitard, which reaches down to approximately 3.20 m bls. Aggregations and mottles of secondary Mn- and Fe-(oxyhydr)oxides mark here the fluctuation range of the unsaturated zone, while permanently reducing conditions are indicated by grey coloured sands in below 3.20 m bls. Where aggregations and mottles of Mn- and Fe-(oxyhydr)oxides occur, trace elements like Zn, Ni, Cu and As are enriched (Figure 6.2). Similar to the sandy aquifer sediments below, As is predominantly strongly adsorbed, but the average content (9.4 mg kg^{-1}) is about three times higher here. In contrast to the assumptions of HARVEY et al. (2006), these sediments appear to act rather as a sink then a source for As. As soon as anoxic groundwater rises into the aerobic vadose zone via capillary rise, the fine grain size of these layers promotes the precipitation of secondary Mn- and Fe-(oxyhydr)oxides, which is in turn accompanied by co-precipitation and/or adsorption of various trace elements, including As. Results from the high As site further support this assumption (see chapter 7.3.2).

6.3.3 ARSENIC IN SEDIMENTS

To assess the As release potential of the aquifer sediments, a multiple physico-chemical approach was used. Analytical limitations arose from the throughout low As contents in the sediments and of the hosting mineral phases. Generally, sedimentary As appears to be closely linked to the fine grain size fraction of the sediments, as suggested by KINNIBURGH & SMEDLEY (2001). Strong positive correlations between bulk As contents and Fe as well as other typically iron-associated trace elements (Cu, Ni and Zn) allude to the conclusion that Fe-(oxyhydr)oxides are important As hosts (see chapter 6.2.1). This assumption is further supported by results of the SEP targeting the identification of arsenic-bearing phases. Here, the biggest part of As turned out to be PO_4^{3-}-extractable, which is interpreted as strongly adsorbed fraction. Remaining sedimentary As is primary associated with Fe-(oxyhydr)oxides (fractions III, IV and V). Significant amounts of HCl-extractable As (fraction III) are not considered to originate from Mn-oxides, since extractable amounts of Mn were negligible compared to Fe (APPENDIX II, Table A 2.2).

According to the high positive correlations of the bulk As contents with Fe and the generally high binding affinities of Fe-(oxyhydr)oxides for As (MOHAN & PITTMAN 2007), the conclusion is drawn that Fe-(oxyhydr)oxides are the dominating host for strongly adsorbed (fraction II) as well as for incorporated As (fractions III, IV and V) in sediments of the low As site. Similar outcomes where reported from the Chakdah site (CHARLET et al. 2007).

6.3.4   GROUNDWATER PROPERTIES

Contradictory arsenic concentrations. Arsenic concentrations in samples of the five monitoring wells were found to be surprisingly much higher than in the adjacent tube well no. 125, which was the reason to choose this location as low As study site (see chapter 5.3.2). Despite comparable depths of the well screens, some important parameters (Cl^-, SO_4^{2-}, NO_3^-, As, Fe, and PO_4^{3-}) strongly differed in values observed in groundwater from well 125. Consequently, well 125 was re-sampled during the field trip in 2009. In this new sample, As, Fe and PO_4^{3-} concentrations were more alike the monitoring wells than to the sample taken during the post-monsoon season in 2007 (Table 6.5). These concentration changes are attributed to on-going redox processes in the aquifer as discussed later on in chapter 6.3.6. Since As concentrations in all five monitoring wells turned out to be throughout high, the designation "low As site" appears inappropriate and the site should be considered rather as a "lower As site" compared to the second study site, where As concentrations are twice as high (see Table 8.1 in chapter 8).

Arsenic release and groundwater evolution. Compared to the field survey (chapter 5.2), numerous additional parameters were determined during the monitoring, allowing a much more precise identification of processes related to groundwater evolution. Solute concentrations, calculated SI and results of the sediment analyses were used to identify hydrogeochemical reactions that have influenced the groundwater composition in the study area. In addition, conservative tracers (Cl^-, K, stable isotope compositions of O and H) were used to identify influences of physical processes like mixing or evaporation (SENGUPTA & SARKA 2006). Nevertheless, subsequent reactions caused by microbial activity or ion exchange reactions may have re-changed distribution patterns of non-conservative solutes like As in groundwater as discussed in the following.

The following processes and mechanisms were identified:

(a) Carbonate and silicate weathering

As previously assumed from results of the field survey, throughout elevated concentrations of Ca, Mg and HCO_3^- in groundwater indicate pronounced influences of carbonate weathering during evolution of the groundwater (Tables 6.4 and 6.7). This is attributed, on the one hand, to the presence of considerable amounts of carbonates in the aquifer sediments (calcite and dolomite), and on the other hand, to the slow groundwater flow supporting the carbonic acid–bicarbonate–carbonate–equilibrium to balance. Dissolved Si concentrations, as well as the mineral inventory, both point at silicate weathering, which is most likely related to the dissolution of instable feldspars (e.g., anorthite).

Due to the carbonate-rich sediments, the groundwater pH is circum-neutral. This controls the protonation of As species and influences in turn their adsorption behaviour. The predominating As species is here uncharged H_3AsO_3 (see E_H-pH-diagram in Figure 6.9). This strongly reduces the importance of ionic surface adsorption / desorption processes, but increases the sorption potential of As(III) to Fe-(oxyhydr)oxides by ligand exchange (DIXIT & HERING 2003). Results of the SEP support these assumptions. Nearly no weakly bound As was observed (fraction I), but strongly adsorbed As (fraction II) was identified as a principal form (even if this fraction was overestimated, see chapter 4.2).

(b) Ion exchange reactions

The presence of surface reactive minerals like Fe-(oxyhydr)oxides and clay minerals generally foster ion exchange reactions, especially Na-Ca-exchange. The Na content of the sediments originates from past transgression events (GOODBRED et al. 2003), when intrusions of marine water caused the reverse exchange reactions (APPELO & POSTMA 1996). Thus, Na/Cl ratios in solution are a good proxy to estimate the groundwater maturity and further support identification of mixing influences, which is used in the following chapters. In contrast to Na-Ca-exchange, PO_4^{3-} to As ratios in solution seem to be not high enough to induce As-PO_4^{3-}-exchange and release of adsorbed As by PO_4^{3-} as proposed by ARCHARYA et al. (1999). This assumption is further supported by results of the in-situ experiments that were carried out the high As and the low As study sites. Here, strong increases in local PO_4^{3-} concentrations did not provoke an additional As release. The long-term monitoring additionally revealed that As to PO_4^{3-} ratios and concentrations in almost all monitoring wells remained constant during the time of observation (see chapter 6.3.6 and chapter 7.3.5.2).

(c) Hydrochemical stratification

Results reveal the presence of two distinctive hydrochemical layers with partially sharp concentrations gradients in many different solutes between the well screens of wells A and B. This set of solutes comprises a wide range of hydro- and geochemically deviant acting compounds including As (Table 6.3). For example, Cl^- is considered a conservative parameter that does not tend to react with solid phase compounds (SENGUPTA & SARKA 2006). In contrast, Na and Ca can be influenced by ion exchange, while As, NH_4^+, SO_4^{2-}, and Co are sensitive to redox changes. The observed stratification effects can be hardly explained by a single process and are attributed to multiple processes. First, groundwaters with different compositions (especially concerning major solutes, like the conservative Cl^-) must have been mixed around the depth range of well A. After mixing, consecutive reactions additionally superimposed the initial distribution of non-conservative solutes including As, what is discussed in the following.

(d) Microbially mediated reactions

Temperatures of around 27°C as observed in the investigation area provide optimum conditions for microbiological growth and activity. Depth decreasing DOC and SOM, as well as the presence of considerable amounts of cultivatable germs and an increased alkalinity in groundwater all indicate pronounced microbial degradation processes in the recent past. Anaerobic decomposition of OM is known to cause formation of fatty acids as intermediate products and HCO_3^- as end product (LOVLEY 1993). Microbial released organic acids further trigger carbonate dissolution through acid neutralisation reactions. Hence, microbial decomposition of OM in reducing aquifer parts entails an increase in groundwater alkalinity, which was demonstrated by the sucrose injection experiment (chapter 6.3.5). In addition, high concentrations of Mn, Fe(II), As(III) coupled with low O_2, NO_3^-, and SO_4^{2-} concentrations reflected a decisive influence of microbial mediated redox reactions on the local groundwater composition. According to the redox classification criteria of JURGENS et al. (2009), methanogenesis is currently the predominating redox process (Table 6.9). Although not being the dominating TEA consuming process any more, Fe(III) reduction can still be a relevant process (JURGENS et al. 2009), which could be demonstrated by the biostimulation experiment (described in chapter 6.3.5).

Table 6.9: Biogeochemical fingerprints of groundwater samples (03/12/09) following the classification scheme of JURGENS et al. (2009). Groundwater from the monitoring wells reflects strongly reducing conditions allowing formation of methane (methanogenesis).

Parameter	NO_3^- (mg/L)	Mn (mg/L)	Fe (mg/L)	SO_4^{2-} (mg/L)
Redox classification criteria:				
Fe(III)/SO_4^{2-} reduction	<2.22	>0.05	>0.10	>0.90
Methanogenesis	<2.22	>0.05	>0.10	<0.90
Concentration in groundwater (depth m bls):				
A (12-21)	<0.88	0.77	4.34	<0.85
B (24-27)	<0.88	0.47	3.76	<0.85
C (30-33)	<0.88	0.41	5.57	<0.85
D (36-39)	<0.88	0.42	2.86	<0.85
E (42-45)	<0.88	0.60	2.15	<0.85

In addition to the problematic enrichment of As, Mn concentrations exceeded the Indian threshold value of 0.10 mg L^{-1} in drinking water (IS 10500; standard value is extendable to 0.30 mg L^{-1} if no other sources of drinking water are available), raising the risk of chronic intoxications for the local population (MC ARTHUR et al. 2012).

(e) Arsenic mobilisation

Similar to arsenic-rich tube wells, hydrochemical properties of groundwater at the low As site appear to be closely related to the microbial reduction of both Fe(III) and As(V). Poorly ordered Fe-(oxyhydr)oxides like ferrihydrite are highly instable under the prevailing hydrochemical conditions, while ordered Fe-minerals (e.g., hematite, magnetite and siderite) are super-saturated and therefore considered as stable (STUMM & MORGAN 1996). From a kinetic point of view, a strong negative or positive SI does not necessarily mean that the respective minerals dissolve or precipitate in a certain period of time (MERKEL & PLANER-FRIEDRICH 2008). However, the occurrence of arsenic-bearing, poorly crystalline Fe-(oxyhydr)oxides in the aquifer sediments (see chapter 6.3.3) can support on-going microbial Fe(III) reduction with concomitant release of As if degradable OM is available. In fact, Fe(III) reduction and As release can be still active processes, which was demonstrated by both, the biostimulation experiment and the long-term monitoring (chapter 6.3.5).

Increased concentrations of Mn reflect that reductive dissolution of Mn-oxides took also place, which often accompanies Fe(III) reduction (LOVLEY 1993). Although aqueous to solid phase ratios of Mn were comparable to those of Fe, concentrations in groundwater and sedimentary contents underline that As mobilisation via Mn(IV) reduction is secondary compared to Fe(III) reduction.

Another potential As release mechanism is biogenic weathering of apatite (MAILLOUX et al. 2009). The presence of potentially arsenic-hosting apatite could neither be proved nor ruled out by the characterisation of the sediments. It is difficult to detect traces of apatite in carbonate-rich sandy sediments by means of XRD, since the main peaks are superimposed by those of quartz. Occurrence of arsenic-bearing apatite was not reported from other study sites in the Bengal Basin until now. According to MAILLOUX et al. (2009), this kind of release would be decoupled from redox processes. Hence, increased As would also appear under less reducing or even oxic conditions, what stands in contrast to common observations in the Bengal Basin.

(f) Competitive adsorption

Groundwater from the investigation area reflects a close connection between As and PO_4^{3-}, which was reported from other locations in the BDP as well (e.g., BGS & DPHE 2001). In the present thesis, the hypothesis was developed that dissolved As and PO_4^{3-} competitively adsorb to residual and newly formed Fe-minerals. This process is considered as an important process in the development of reducing groundwaters in the Bengal Basin as well as other Asian delta areas.

Reductive dissolution of Mn- and Fe-(oxyhydr)oxides should be accompanied by a fast re-adsorption of previously released As until the sorption capacity is reached or available potential sorbents are completely dissolved (DIXIT & HERING 2006, WELCH et al. 2000). Geochemical data support the former, since amorphous and weakly ordered Fe-(oxyhydr)oxides with considerable amounts of As are still available within the aquifer sediments, despite the strong negative SI (see chapter 6.3.3). In contrast to the prevailing reducing conditions, some Fe-mineral phases that are capable of arsenic retention are supersaturated in groundwater (see SI of hematite, magnetite and siderite in Table 6.4). While precipitation of Fe-(oxyhydr)oxides like hematite or magnetite is considered unlikely due to the nearly absence of dissolved O_2, previous studies revealed that high concentrations of dissolved Fe(II) can induce transformation of weekly ordered Fe-(oxyhydr)oxides into more stable forms like magnetite. These transformed minerals are again capable of As(III) and As(V) adsorption (HANSEL et al. 2003, PEDERSEN et al. 2005) as well as incorporation of previously adsorbed As(V) into the magnetite crystal structure (COKER et al. 2006). In addition, precipitation of siderite is also possible (positive SI values according to the low redox potential and the throughout high HCO_3^- and Fe concentrations), which has a pH-E_H-stability field close to magnetite and is able to remove dissolved As(III) as well as As(V) by adsorption (GUO et al. 2007, JÖNSSON & SHERMAN 2008). Detection of siderite in the sediment samples by XRD was not possible (principal siderite peaks are superimposed by mica and feldspar peaks), but magnetite was proved by bar magnet. However, magnetite and concretes of siderite have been previously identified in sediments of West Bengal under reducing conditions (PAL et al. 2002).

In fact, outcomes of the field survey and the biostimulation experiment strongly support the assumption of As (re-)adsorption within aquifers of the investigation area (see chapters 5.3.1 and 6.3.5). The predominating As species in local groundwater is always As(III), which can strongly adsorb to free binding sites of Fe-mineral surfaces via inner-sphere surface complexes (ONA-NGUEMA et al. 2005). At neutral pH values, PO_4^{3-} has higher binding affinities for Fe-minerals than both, As(III) and As(V) (GOH & LIM 2004). Hence, it can be expected that adsorption to newly formed Fe-minerals is competitive and that PO_4^{3-} adsorption is preferred. All available As and PO_4^{3-} binding sites in the aquifer sediments are considered to be currently occupied because of the constantly increased concentrations in groundwater. This assumption is further supported by the high percentages of strongly adsorbed As in local sediments, based on the SEP results. The aqueous to solid phase ratios (Table 6.3) show that As is relatively more enriched in groundwater than PO_4^{3-}, suggesting that PO_4^{3-} is favoured over As(III) and/or As(V) adsorption. Depending on available binding sites and prevailing solute concentrations, distinctive As-PO_4^{3-} ratios developed in groundwater (Table 6.10). This alludes to the conclusion that As concentrations in groundwater would be distinctively lower in absence of dissolved PO_4^{3-}. The concept of competitive adsorption of As and PO_4^{3-} is further developed in the following chapters.

Table 6.10: Comparison of As to PO_4^{3-} mol ratios in groundwater (samples from 03/12/09) and in sediments from respective depths (determined from micro-wave acid digestions).

Well (depth m bls)	Groundwater (As to PO_4^{3-} mol ratio)	Sediment (As to PO_4^{3-} mol ratio)
A (12-21)	1 : 37	1 : 268
B (24-27)	1 : 17	1 : 237
C (30-33)	1 : 6	1 : 300
D (36-39)	1 : 13	1 : 312
E (42-45)	1 : 11	1 : 127

(g) Anthropogenic influences

Since the study site is located in a rural area of intense agriculture, the question arises whether there are any traceable anthropogenic influences that influence the mobilisation and distribution of As in the groundwater. Three possible effects were identified:

- Inflow of anthropogenic derived OM

    An intensively discussed issue is the infiltration of OM either by sewage from pit latrines, by artificial ponds, or by irrigation water from fertilised fields. Since OM is considered as a limiting factor for microbial mediated reactions in aquifer sediments of the BDP (RADLOFF et al. 2008), an inflow of OM into the aquifer would most likely induce additional As release (NEUMANN et al. 2009). However, DOC concentrations in the monitoring wells range below 10 mg L^{-1}, which is considered as common for pristine groundwater (COZZARELLI & WEISS 2007). Total microbial abundance as indicated by TPC reflects a moderate burden of the groundwater, which is fostered by the comparatively high groundwater temperatures of about 27°C.

The absence of NO_3^- and coliform bacteria (especially *E. coli*) further supports the assumption that groundwater at the study site is not affected by sewage or organic fertiliser.

- Influences of excessive pumping

  Potential influences on the groundwater composition can arise from an adjacent irrigation pump. Since irrigation is restricted to the dry pre-monsoon season, this issue is discussed in context of the monitoring results (see chapter 6.3.5). Nevertheless, the pumping stations of Chakdah described by NATH et al. (2008) are assumed to have no or minor effects on the local hydrology, since they are located more than 10 km east of the study site.

- Influences of wet rice cultivation

  Cultivation of wet rice is widespread in the BDP (NORRA et al. 2005). The typically clayey paddy field soils constitute an effective barrier that prevents infiltration of O_2 into the aquifer parts below, supporting development of reducing conditions and by association As release (MÉTRAL et al. 2008).

6.3.5 INFLUENCE OF THE AVAILABILITY OF ORGANIC MATTER ON THE RELEASE OF ARSENIC: THE BIOSTIMUALTION EXPERIMENT

Microbial metabolic pathways can be directly identified and quantified only with latest cutting edge methods of molecular biology, which enable to determine the share of active microbial species based on the present DNA, respectively RNA (FREEMAN & GOLDHABER 2011). Unfortunately, such kind of analyses exceeded the analytical possibilities of the project. Hence, conclusions regarding active microorganisms need to be indirectly drawn from concentration changes of available TEA and released metabolic products such as fatty acids.

Where possible, interpretations of the results obtained from this experiment are compared to similar experiments (e.g., HARVEY et al. 2002). Available literature is generally limited, in particular regarding field experiments (RADLOFF et al. 2008). Additionally, hydrochemical analyses done for laboratory incubation experiments are often restricted to As and Fe and do not consider other trace elements.

Effects of pumping during sucrose injection. It is necessary to distinguish between effects related to the introduction of sucrose via circular pumping and alterations arising from microbially mediated consecutive reactions. Consequences of abiotic mixing induced by circular pumping can be best observed in the shallowest well A, in which no sucrose was directly added and therefore nearly no sucrose occurred. Since initial baseline values of groundwater in well A differed significantly from the other wells, concentration changes for example in conservative Cl^- in samples taken two days after sucrose addition by circular pumping reflected mixing influences that were visible in all wells. Additionally, As and Co concentrations behaved inversely. Both slightly increased in well A and concomitantly decreased in wells B and E.

Phosphate behaved differently from all other parameters. Direct increases in wells C and E following circular pumping were several times higher than the respective baseline values. This behaviour is attributed to entrainment of loosely bound, electro-statically adsorbed PO_4^{3-} during circular pumping (DORIOZ et al. 1989, REDDY et al. 1999). A similar effect occurred during the groundwater abstraction experiment at the high As site, where even higher increases occurred (chapter 7.3.5.1). Despite this, PO_4^{3-} concentrations rose in wells A, B and D, too, but with a lag of two days. A relation to microbial processes is considered unlikely to this early point of time. This is supported by the control well A, where a similar PO_4^{3-} increase occurred despite the nearly absence of sucrose.

Surprisingly, no sucrose was detectable in samples from well C taken two days after sucrose insertion, but four days later. As a result, subsequent changes due to microbial decomposition of sucrose appeared to be delayed as well, as discussed in the following.

Availability of sucrose determines the intensity of microbial reactions. Changes in the hydrochemical composition of samples have shown to directly depend on the availability of dissolved sucrose. Groundwater of monitoring well E had the highest initial sucrose concentrations and showed strongest influences of microbially mediated metabolic reactions. Similar changes in groundwater chemistry occurred also in wells B, C and D, but intensities were less and partly delayed (well D) due to the lower sucrose availability. Due to the low sucrose concentrations, nearly no changes occurred in the chemistry of well A. In spite of the low sucrose input, microbial growth was induced, as indicated by slightly increasing alkalinity and TPC values. Changes in As and Cl$^-$ are solely attributed to circular pumping. Hence, well A is considered to be suitable to control changes induced by sucrose addition.

The quick response of indigenous microbial populations as indicated by an escalating TPC proved the presence of microbes in the aquifer, which were successfully stimulated via sucrose addition. Further on, circular pumping proved to be a suitable method to introduce dissolved substances into an anaerobic system without introducing external O_2, which is toxic to most strictly anaerobic microbes (KIEFT et al. 2007). Compared to preliminary column experiments conducted by D. FREIKOWSKI, in-situ reactions have been proved to be even faster and more intensive. Very likely, the relatively high groundwater temperature of around 27°C provided optimum conditions for microbial growth and activity, much similar to laboratory incubation experiments.

Microbial reactions following biostimulation. Groundwater samples were taken every 48 h for the following two weeks to monitor changes in the local hydrochemistry following sucrose introduction. The recorded trends reflected an efficient stimulation of cooperative anaerobic microbes in wells B, C, D and E. Rapid microbial growth was indicated by an exponential rising number of free germs (TPC), which was accompanied by formation and subsequent consumption of intermediate catabolic products such as fatty acids (predominantly acetate). The strong increase in dissolved Fe(II) concentrations reflects that dissimilatory Fe(III) reduction was involved in the degradation of sucrose. This assumption is further supported by a positive correlation between Fe(II) and acetate (Figure 6.16, A). Acetate, which is considered the most important fermentation product related to microbial Fe(III) reduction, has occurred two to six days after infusion of sucrose (LOVLEY 1993, LOVLEY et al. 2004). The formation of considerable amounts of fatty acids was accompanied by declining pH values, especially in well E, in which the strongest reactions could be observed. With an approximate delay of two days relative to the occurrence of acetate, the activity of FeRB was indicated by strongly increasing Fe(II) concentrations. Manganese was released parallel to Fe and formed similar trends, but with less absolute and relative increases. Hence, microbes using Mn(IV) reduction to decompose OM were likely active, too, but with a far lesser impact compared to Fe(III) reduction.

Concentration increases in dissolved Fe(II) appeared as throughout too low as compared to the peaking acetate concentrations. For example, maximum acetate concentrations in well E were 7.66 mM (12/12/09). If this amount was solely ascribed to the degradation by *Geobacter metallireducens* (a typical FeRB), the concentration of Fe(II) in groundwater would have increased by 30.6 mM (according to equation 2.1, chapter 2.2.3). Instead, the maximum increase in dissolved Fe in this well was only 1.35 mM. The same is true for metabolically released HCO_3^-. According to equation 2.1, the decomposition of one mol acetate should result in a HCO_3^- to Fe(II) ratio of 1 : 4. Instead, results for well E (12/12/09) display an approximate ratio of 1 : 0.45 (Figure 6.16, B).

Hence, acetate was consumed and HCO_3^- must have been released by other anaerobic microbes in addition to FeRB (e.g., methane producing microbes during methanogenesis).

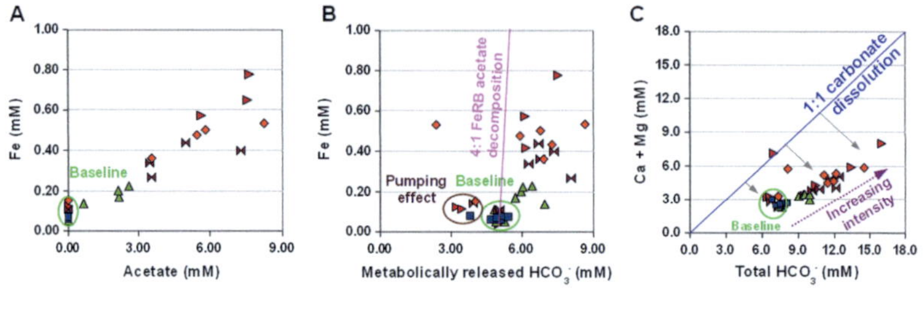

Figure 6.16: A) Positive correlations between Fe and acetate concentrations after sucrose addition (samples 03/12/09 to 18/12/09). B) Both, Fe and metabolically released HCO$_3^-$ increased, but never reached stoichiometric ratios that are characteristic for pure FeRB cultures. Metabolically released HCO$_3^-$ was calculated as difference of total HCO$_3^-$ and carbonate HCO$_3^-$ mol concentrations (carbonate HCO$_3^-$ = Ca + Mg mol concentrations). C) With on-going microbial decomposition of sucrose, Ca and Mg as well as total HCO$_3^-$ concentrations increased, with increases in HCO$_3^-$ clearly exceeding carbonate dissolution.

Another possibility is a continuous removal of dissolved Fe(II) via precipitation and adsorption. Considering the strongly reducing conditions, when the maximum Fe(II) mobilisation was reached in wells B and E, only two Fe-minerals, siderite ($FeCO_3$) and vivianite ($Fe_3(PO_4)_2$ x $8H_2O$), could have precipitated according to thermodynamic calculations (positive SI, Table 6.7). A quick precipitation of siderite has been previously linked to Fe(III) reduction in systems with increased alkalinity (KONHAUSER 2007). ISLAM et al. (2005) further described removal of As(V) and As(III) from solution by sorption to biogenic siderite and vivianite during a three week long incubation experiment conducted with *Geobacter sulfurreducens*, a Fe(III)-reducing bacteria. As discussed in chapter 6.3.4 (f), under reducing conditions and in absence of H_2S, high concentrations of dissolved Fe(II) can further entail the transformation of residual Fe-(oxyhydr)oxides into more stable forms, which is primarily magnetite (e.g., O'LOUGHLIN et al. 2010, YANINA & ROSSO 2008).

Since microbial communities are highly diversified and groundwater is strongly reducing and enriched in reactive solutes like Fe(II) and HCO_3^-, it is most likely that precipitation of siderite and transformation of residual weakly ordered Fe-phases into magnetite occurred more or less successively in course of the experiment as a function of the prevailing redox potential and pH.

Consecutive reactions. The microbial degradation of sucrose to fatty acids and finally CO_2/HCO_3^- caused an increase in alkalinity and release of protons (e.g., KONHAUSER et al. 2011, LOVLEY & PHILLIPS 1989). Hence, alkalinity increased by metabolic reactions as well as by carbonate dissolution following acid neutralisation reactions (Figure 6.16, C). Importantly, throughout increased alkalinities in the baseline values (before sucrose was introduced) clearly point at a high microbial activity during previous groundwater evolution as shown in chapter 6.3.4 (d).

Since most trace metals turned out to be primarily associated with Fe-(oxyhydr)oxides, increases in Ni, Zn, Co, V and As concentrations are primary attributed to Fe(III) reduction. Temporary low pH values have further fostered the mobility of adsorbed trace elements like Zn and Ni (CAPPUYNS & SWENNEN 2008). Changes in the redox state of adsorbed trace metals has likely influenced their mobility, too. For example, reduction can increase the mobility of As, Zn, Ni and Co, while decreasing that of U, V, and Mo (ADRIANO 2001, WEHRLI & STUMM 1988). In addition, organic acids and HCO_3^- can form complexes with cations, increasing their mobility and preventing adsorption and co-precipitation (BORCH et al. 2010). Hence, multiple and partly counteracting mechanism may have occurred after sucrose addition, significantly influencing the mobility of certain trace elements. The pool of elements which can be mobilised is generally a function of the prevailing mineralogical composition of the aquifer sediments, which is homogenous in range of the well screens (chapter 6.3.1). Hence, differences in the mobility of major and trace elements were obviously solely controlled by geo-microbiological factors, which were in turn controlled by the availability of e^--donors in form of sucrose and its decomposition products.

Return to baseline values. After the complete consumption of sucrose and its intermediate catabolic products, microbial growth and also metal reduction stopped, as indicated by stagnation of the TPC and the releases of Fe and Mn. Calcium and most trace elements gradually returned towards the initially recorded baseline values. Strongly increased parameters (e.g., TPC, Fe and PO_4^{3-}) stagnated at first, or declined very slow. Declining trends of major elements are primarily attributed to precipitation of super-saturated mineral phases, while trace elements have been rather adsorbed by residual minerals and/or by newly formed minerals (ZACHARA et al. 2001). In addition, mixing with slowly inflowing unaffected groundwater appeared as indicated by trends of conservative Cl^-.

Trends in well D appeared delayed relative to the other wells, since none of the increasing parameters (except for As and PO_4^{3-}) peaked during the two weeks of detailed sampling. The subsequent monitoring revealed that the "missing" peaks occurred before 06/01/10, when the next monitoring sample was taken.

Concomitant mobilisation and retention of arsenic. Total arsenic and As(III) concentrations in groundwater of the sucrose influenced wells increased during the experiment, but relative increases of up to 49 % (well E, As) were throughout low compared to the baseline values. The high As concentrations are considered to result primarily from microbial Fe(III) reduction, since Fe-(oxyhydr)oxides were identified as principal As hosts in the investigated sediments and considerable amounts have been dissolved after sucrose addition. The close relationship between As and Fe is displayed in Figure 6.17. Despite this, potential influences of Mn(IV)-reducing microbes on the release of As are considered marginal due to the throughout low increase in the Mn concentrations. Since As(III) was found to be constantly the dominating As species, activity of As(V)-reducing bacteria cannot be excluded, especially since many FeRB carry the necessary gene sequences (OREMLAND & STOLZ 2005, ZOBRIST et al. 2000). Cultivation experiments conducted by CAMPBELL et al. (2006) with *Shewanella* sp. strains demonstrated that As(V) reduction via the detoxification pathway required relatively high As(V) concentrations (>100 µmol L^{-1}), whereas the *arrA*-gene necessary for the metabolic pathway was expressed at comparatively low concentrations (>100 nM). Hence, the activity of DARPs cannot be excluded for this experiment. Figure 6.18 provides a summary of the processes and mechanisms involved in the biostimulation experiment in form of a conceptual model.

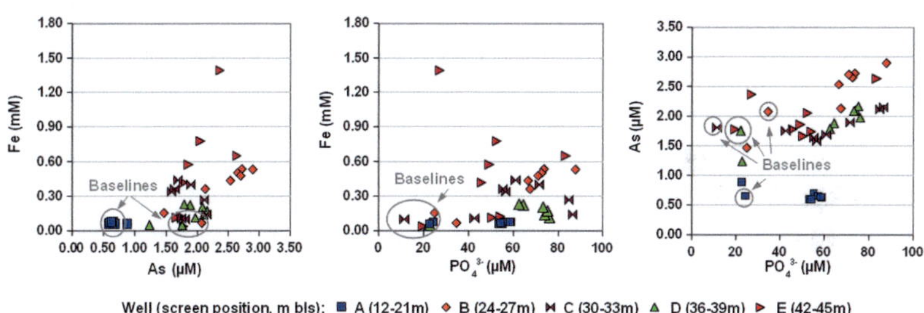

Figure 6.17: Dissolved concentrations of Fe, As and PO$_4^{3-}$ (03/12/09 – 18/12/09) display a close relationship between the three parameters in wells B, C and E. In addition, the pronounced shift in the PO$_4^{3-}$ concentrations following circular pumping is clearly visible in the plots. Respective baseline values (03/12/09) in the plots are marked by grey circles.

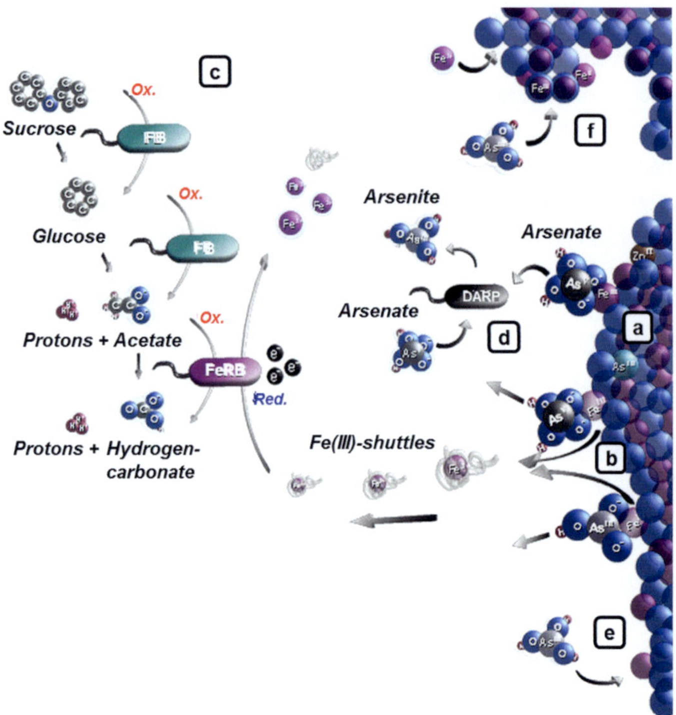

Figure 6.18: Arsenic mobilisation and retention during the experiment.

a) Simplified scheme illustrating the reductive dissolution of ferrihydrite that hosts As(V) and As(III). b) Complexation and mobilisation of Fe(III) by Fe(III)-shuttles triggers the passive release of adsorbed and incorporated trace elements including As(III) and As(V).   c) Sequential dissimilatory decomposition of sucrose by a consortium of cooperating microbes with different metabolic pathways (fermenter, FeRB, DARPs) that causes the release of intermediate catabolic products, including protons, fatty acids and reduced Fe(II). d) Potential dissimilatory reduction of As(V) by DARPs. e) Fast re-adsorption of previously mobilised As(III) and/or As(V) onto free binding sites of residual ferrihydrite.   f) High concentrations of Fe(II) trigger transformations of residual ferrihydrite, resulting in the formation of magnetite and adsorption of As(III) and/or As(V). Similar behaviour is expected for PO_4^{3-}, which is additionally released from sucrose decomposition and competed with As(III) and As(V) for binding sites during adsorption.

Concentrations of other trace elements (Zn Ni, Co and V) increased much more compared to As, especially in the deepest well E, where most intensive microbial reactions occurred. These elements are primarily associated with sedimentary Fe-(oxyhydr)oxides according to sediment analysis. This stronger mobilisation can be explained by additional release mechanisms, like pH controlled desorption or changes in the redox state. Nevertheless, As mobilisation appears to be at least in part inhibited, since the net increase in As was generally decoupled from the initial sucrose concentrations and the intensity of Fe(III) reduction. The gap between the amount of released As and Fe(II) increased with time, until an As to Fe mol ratio of approximately 1:100 was reached four to eight days after sucrose insertion in all wells (Figure 6.10). Such a decoupling was predicted by HERBEL & FENDORF (2006), PEDERSEN et al. (2006) and WELCH et al. (2000) for natural aquifer systems in state of Fe(III) reduction.

According to the geochemical characteristics of the sediments, this decoupling cannot be attributed to a limited availability of sedimentary As. Hence, fast re-adsorption of the mobilised As by residual Fe-(oxyhydr)oxides as well as a subsequent retention by newly formed Fe-minerals can be assumed. When the redox potential temporarily strongly declined and Fe(II) concentrations rose, it can be assumed that first magnetite was formed by transformation of residual, less stable iron phases through dissolved Fe. Where pH and redox potentials reached the necessary low values, siderite and even vivianite could have precipitated. For example, when strongest reactions occurred, magnetite formation was thermodynamically favoured in range of well D, siderite precipitation in case of well B, and vivianite precipitation in well E (see Table 6.7).

The (re-)adsorption of As also explains the observed decoupling of As mobilisation from Fe release. Other possibilities of As retention were excluded as far as possible. Immobilisation of As and other trace elements due to formation of sulphides was excluded, as there was only marginal dissolved SO_4^{2-} available. The possibility of volatile arsine formation by anaerobic microbes cannot be ruled out, but it is considered as unlikely, since detoxification mechanism require much higher As concentrations to be activated (CAMPBELL et al. 2006). To estimate the possibility of As loss in filtrated samples by arsenic-trapping Fe-colloids that exceeded the filter size of 0.45 µm (GUO et al. 2009), the As concentration in filtered and unfiltered samples was compared. Observed differences remained <3 %, indicating that no loss of As appeared by filtration of groundwater samples.

Mid-term effects following sucrose injection. The successful stimulation of indigenous FeRB by sucrose has caused a pronounced temporary change in the prevailing As-PO_4^{3-}-Fe-ratios of the groundwater. Once available OM was exhausted, precipitation of supersaturated mineral phases dominated, and the system returned to a state close to hydrochemical equilibrium. Figure 6.19 summarises the described processes during and after the biostimulation experiment in form of a conceptual model. Here, it can be distinguished between a fast re-adsorption that accompanied directly the release of As and PO_4^{3-} (Figure 6.19, B) and a subsequent adsorption related to precipitation and formation of new Fe-minerals (Figure 6.19, C and D). A slow lateral inflow of unaffected groundwater followed the monsoon rains end of April. At this time, declines stopped suddenly and parameters stabilised all of a sudden.

Although parameters returned to near-baseline concentrations, Fe and As remained at rather elevated levels in the sucrose influenced wells until the end of the groundwater monitoring in August 2010. Hence, sucrose addition presumably caused a net loss of available As binding sites, which was not (yet) compensated by precipitation of new Fe-minerals.

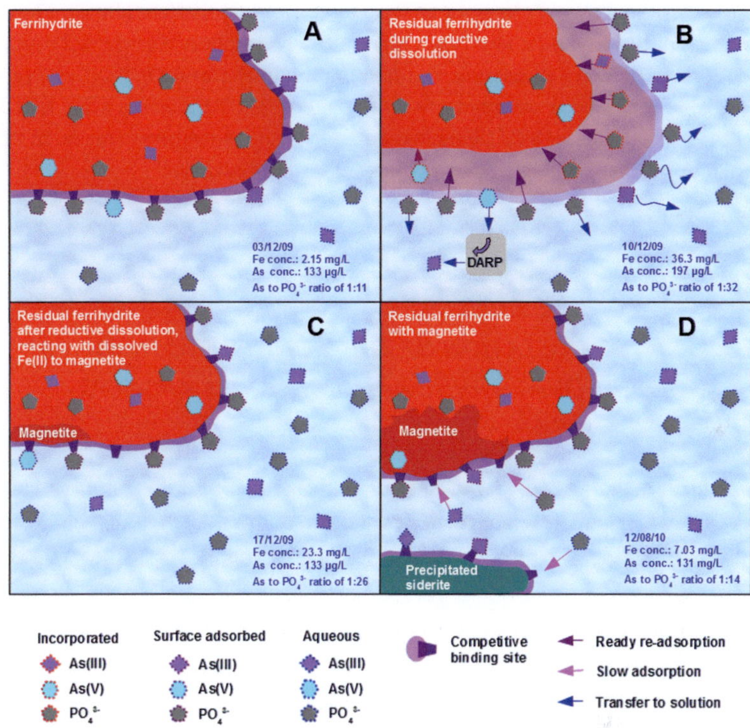

Figure 6.19: A) Initial situation before the sucrose introduction, where Fe(III) reduction has stopped or slowed down after the pool of degradable OM was exhausted and available As and PO_4^{3-} binding sites were occupied. B) and C) Addition of sucrose fuelled new Fe(III) reduction, causing the release of incorporated and adsorbed As and PO_4^{3-}, and thereby their fast competitive re-adsorption on residual Fe-(oxyhydr)oxides and newly formed Fe-minerals. After sucrose was exhausted, As release stopped, whereas As and PO_4^{3-} concentrations in groundwater remained high due to a net loss of binding sites. D) Since high amounts of Fe(II) were dissolved, transformation and precipitation of new Fe-phases continued in the following months. Figure caption continued on next page. Figure caption continued on next page.

Figure 6.19, continued: New As binding sites were generated, allowing further removal of dissolved As and PO_4^{3-}. Since PO_4^{3-} adsorption is preferred over As, adsorption caused a subsequent shift in the groundwater As to PO_4^{3-} mol ratios in favour of As. Not included are DARPs and specific binding-sites that are restricted to As(III) or PO_4^{3-}. Included Fe and As concentrations and As to PO_4^{3-} mol ratios refer to well E, where strongest reactions occurred.

6.3.6 *INTERPRETATION OF THE MONITORING RESULTS: SEASONAL TRENDS AND ONGOING ARSENIC RELEASE*

The question arises, what kind of processes control the mobilisation of As and which are still active. Evidence for current As release coupled to intensive microbial degradation was previously reported from the Chakdah area, where long-chained hydrocarbons are still subject to slow, but consequent decomposition through FeRB as demonstrated by microbial incubation experiments (ROWLAND et al. 2006 & 2009). Since residual Fe-minerals contain significant amounts of adsorbed and incorporated As, on-going activity of FeRB and/or DARPs should further induce As mobilisation at the low As site by reducing available binding sites.

Seasonal effect. Although As and PO_4^{3-} concentrations are generally controlled by biogeochemical processes, a strong correlation of these parameters with the hydrostatic head and conservative elements (e.g., Cl^-, Na) could be observed in monitoring samples from well A. Hydraulic conditions changed periodically from unconfined towards confined, every time the rising water table reached the base of the near-surface aquitard (at 3.20 m bls) during the monsoon season. The close connection between changes in As concentrations, conservative tracers and the hydrostatic head strongly suggests that described trends in 2009 and 2010 were caused by groundwater movement.

Similar temporary changes were previously reported from a shallow monitoring well in Bangladesh and from wells of the JAM study site (DHAR et al. 2008, MC ARTHUR et al. 2010). The hydrogeological situation in this area is quite similar to the low As site: local groundwater is hydrochemically stratified and covered by a clayey aquitard. Changes in the groundwater composition were attributed to vertical and horizontal movement of the local water bodies during the monsoon seasons.

At the low As study site, a direct inflow of surface water (irrigation water, rain water or sewage) is unlikely, since the actual aquifer is covered by an aquitard (discussed in 6.3.2). Nevertheless, the hydrostatic head continuously rose during both monsoon seasons, reflecting recharge connected to the monsoon rains. Rainwater is supposed to have infiltrated via adjacent sandy lenses within the surface aquitard, as suggested by MÉTRAL et al. (2008) for a study site west of Chakdah.

Hence, the study site is considered to be indirectly influenced by infiltrating rain water during the monsoon seasons, which causes displacement and mixing of groundwater. Since the well screens are located in depths below 12 m bls, it is generally difficult to identify such an influence in the monitoring samples. Changes in the groundwater chemistry during the monsoon season allude to the conclusion that percolating water is increased in Cl^- and Na, and depleted in As and PO_4^{3-}. In addition, mixing effects following the strong rainfall in 2009 caused a visible shift in the $\delta^{18}O$ values of about 0.50 ‰ towards the heavier isotope. In contrast, the infiltration of sewage or fertiliser is unlikely, since no input of NO_3^- was detectable during the monsoon season.

This effect is obviously reversible, since opposite trends developed during the dry seasons in 2009 and 2010, when the groundwater table had declined. In 2010, the drop in the water table was greater and it was accompanied by much more pronounced changes in the hydrochemistry. During the dry season, adjacent fields are irrigated by a 24 m deep pumping well, located about 75 m from the study site. Excessive groundwater abstraction has potentially caused redirection and an increase of the local lateral groundwater flow during the dry season. For example, attraction of arsenic-enriched groundwater towards pumping wells was previously described in context of a larger scale for Chakdah (CHARLET et al. 2007) and a pumping field near Hanoi, Vietnam (NORRMAN et al. 2008). Within the following considerations, it is necessary to distinguish between seasonal effects and changes in the hydrochemistry that are related to on-going processes of groundwater evolution.

Evolution of groundwater chemistry. During the monitoring, well A displayed a continuous change towards a hydrochemical composition similar to that in the deeper wells. Concentrations of As, PO_4^{3-} and Fe have multiplied relative to the first sampling in January 2009. These net increases added up to the previously described seasonal variations in As and PO_4^{3-} concentrations, and additionally included Fe (Figure 6.14). Hence, these results reveal that Fe(III) reduction and concomitant As and PO_4^{3-} release is obviously an active process in the shallow aquifer part in around 12 to 21 m. The influence of the biostimulation experiment in this part of the aquifer is considered as negligible, since the sucrose input was marginal and the increase had already began in January 2009. An influence of microbial mediated Mn(IV) reduction is also excluded according to throughout stable Mn concentrations in all wells. Similar changes occurred in the adjacent and similar deep tube well 125, which was first sampled during the field survey in 2007, and two years later again (see Table 6.5, well 125a and 125b). During this time period, As, PO_4^{3-} and Fe concentrations have strongly increased, while NO_3^- disappeared. This trend indicates changes in the local redox state, which has gradually shifted from NO_3^- to Fe(III) reduction.

In the other monitoring wells, As and PO_4^{3-} concentrations had remained at relatively stable levels between April and the sucrose experiment in December 2009, which means that no noteworthy As release or immobilisation occurred here. Arsenic to PO_4^{3-} ratios in bulk sediment range between 1:438 (depth range of well C) to 1:595 (around well E), whereas in groundwater characteristic and relatively stable ratios prevailed since April 2009. The As to PO_4^{3-} ratio in groundwater of the redox active well A (~1:32) was throughout different from the four other wells (between 1:17 and 1:14). The ratio of well A was equal to the value that adjusted in groundwater of well E when As release reached its maximum during the biostimulation experiment (see Figure 6.19, B). These shifts in respective ratios suggest an increasing influence of competitive adsorption once As and PO_4^{3-} release has slowed down or completely stopped, e.g., after the biostimulation experiment (Figure 6.19, C and D).

Before April 2009, $\delta^{18}O$ values, As(III) percentages and PO_4^{3-} concentrations strongly varied in most monitoring samples, which is interpreted as a consequence of drilling of a new central well A in November 2008. According to common practice, the well was intensively "flushed" after completion, thereby causing pronounced perturbations that lasted several weeks before the hydrochemistry returned to the initial baseline values.

Comparison to literature. Available literature presenting results from in-situ experiments related to the mobility of As is very rare. The only two available studies that used in-situ addition of organic matter in the BDP (Bangladesh) to monitor influences on As mobility were conducted by HARVEY et al. 2002 and SAUNDERS et al. 2008. Experimental setups and focuses were different and in both cases molasses was used as OM source. HARVEY et al. (2002) injected molasses into a 31 m deep well with 8.5 µM As, which first caused increasing As concentrations caused by FeRB, followed soon by decrease (attributed to sulphide precipitation). SAUNDERS et al. (2008) first introduced molasses into a shallow single well, followed by $MgSO_4$ after 127 d to stimulate SRB and to immobilise dissolved As via sulphide formation. Here, dissolved As concentrations first declined before they re-increased, which is attributed to a lack of available SO_4^{2-} to compensate As release caused by FeRB.

In both studies, no trace elements other than As and Fe were taken into account. Hence, the comparability with outcomes of the biostimulation experiment done at the low As site is strongly limited and the newly developed concept of competitive As and PO_4^{3-} adsorption cannot be validated. Results only support the assumption that microbiological reactions strongly influence the mobility of As.

The same applies to microbiological column experiment studies using FeRB (e.g., ISLAM et al. 2005, KOCAR et al. 2006, PEDERSEN et al. 2006, TUFANO & FENDORF 2008), which focused on the As-Fe system, but did not consider other trace elements, especially the behaviour of PO_4^{3-}.

In 36 wells located in Bangladesh that were monitored by BGS & DPHE (2001), generally minor changes in As concentrations appeared over time. These changes were neither related to seasonal effects nor on-going Fe(III) reduction. Trends in PO_4^{3-} concentrations generally traced those of dissolved As, further supporting the assumption that both are controlled by similar mechanisms.

The few other hydrochemical monitoring campaigns that were conducted at study sites in West Bengal (MC ARTHUR et al. 2010, NATH et al. 2008) and Bangladesh (VAN GEEN et 2007) described temporary changes in concentrations of As groundwater. In all studies, temporal changes (comprising increases as well as decreases) were attributed to mixing of groundwaters with deviant dissolved As concentrations. The same applies to the Red River Delta in Vietnam, where pumping could be identified as controlling mechanism for increasing As concentrations (WINKEL et al. 2011). In no case, increasing As concentrations could be linked to an active mobilisation as described for well A at the low As site.

7. THE HIGH ARSENIC STUDY SITE

7.1   INTRODUCTION

Investigations at the high As site primarily focused on abiotic influences on mobilisation and distribution of As in groundwater of the BDP (NEIDHARDT et al. 2012c). The approach was generally similar to the low As site. First, the spatiotemporal distribution of As and other solutes was recorded in local groundwater based on bi-weekly samples taken from five nested multilevel wells. After one year, an in-situ experiment was conducted to simulate shallow groundwater abstraction, which is widespread in the Nadia district and the entire BDP (Nath et al. 2008). After that, the monitoring was continued in order to determine potential mid-term effects related to this experiment. The hydrochemical data was completed by a thorough characterisation of the aquifer sediments that were obtained during well drilling to identify As hosts and draw conclusions regarding the aquifer architecture and sedimentation history.

The monitoring wells are located in the backyard of a farming family in the small village Sahispur. The village consists of a few homesteads along an unconsolidated road and is surrounded by agricultural fields. A small pond with approximate dimensions of 7 x 4 m and a depth of about 2 m is located adjacent to the monitoring wells. This offered the opportunity to examine potential connections between infiltrating pond water and the aquifer below, which was assumed by FAROOQ et al. (2010) and NEUMANN et al. (2009).

7.2   RESULTS AND INTERPRETATION

7.2.1   *SEDIMENT CHARACTERISATION*

Stratigraphy and geochemistry. The five well screenings are located in different depths of a single aquifer that is composed of fine to medium grained sand including varying proportions of silt. Similar to the low As site, layers of silty clay and clayey silt with a total thickness of 3.35 m (entitled as facies F4) confine the aquifer towards the surface (Figure 7.1). Buried clinker fragments occur down to a depth of 2.00 m bls, whereas in 2.70 m bls conches of molluscs appear. Between 2.05 to 4.65 m bls, Ca and TIC contents are lowest. In addition, reddish and black aggregations and mottles of Mn- and Fe-(oxyhydr)oxides occur here that are accompanied by highest contents of Fe, Mn, Ni, Cu, Zn, and As (Figure 7.2). Between 2.70 and 5.95 m bls, the sediment colour changes from brown to olive grey and below 5.95 m bls to reduced dark grey. Here, the intensity of visible reactions with HCl decreases from strong to low. Between 3.35 and about 17.0 m bls, the silt content gradually decreases with depth and with it TOC, major and trace elements with the exception of Zr and Ce (facies F3b, Figure 7.2). A sharp gradient of the grain size distribution is located in about 17.0 m bls, where medium sand occurs. Between 17.0 to 32.0 m bls, most major and trace elements reach lowest contents and show relatively invariant depth distributions (facies F3a). In the sediments located between 32.0 to 39.2 m bls (facies F2), several thin layers of poorly sorted sediments are interbedded (ranging from fine gravel to clay), which are accompanied by sporadic changes in the sediment colour (from dark grey to brownish) and partly varying contents of TOC and most major and trace elements. The fine gravel is composed of poorly to medium rounded quartz pebbles and carbonate nodules.

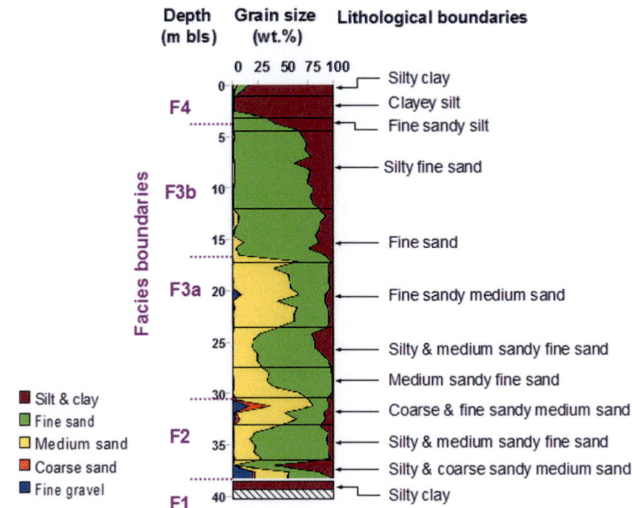

Figure 7.1: Grain size composition in sediment samples and position of the litho-facies boundaries.

In F2, TOC, TS and most trace elements including Fe and As have pronounced peaks similar to the surface near clay layer (Figure 7.2). This part is further underlain by a clay layer beginning in 39.2 m bls (facies F1), which is why drilling was stopped before the intended final depth of 45.0 m bls was reached.

Sandy sediments located between 3.35 and 39.2 m bls (facies F2, F3a and F3b) form one shallow aquifer, which is confined by clayey aquitard layers towards the bottom and the surface. Hydraulic conductivities (K) were estimated with the Beyer formula (BEYER 1964) based on the grain size distribution. Respective K values range from 2.6×10^{-8} m sec^{-1} (poorly permeable) in case of the clayey layers, over 4.4×10^{-5} m sec^{-1} (permeable) for the silty fine sands, and up to 3.6×10^{-4} m sec^{-1} (well permeable) for the medium sand rich strata. Complete results are summarised in APPENDIX III, Table A 3.1 and Figure A 3.4.

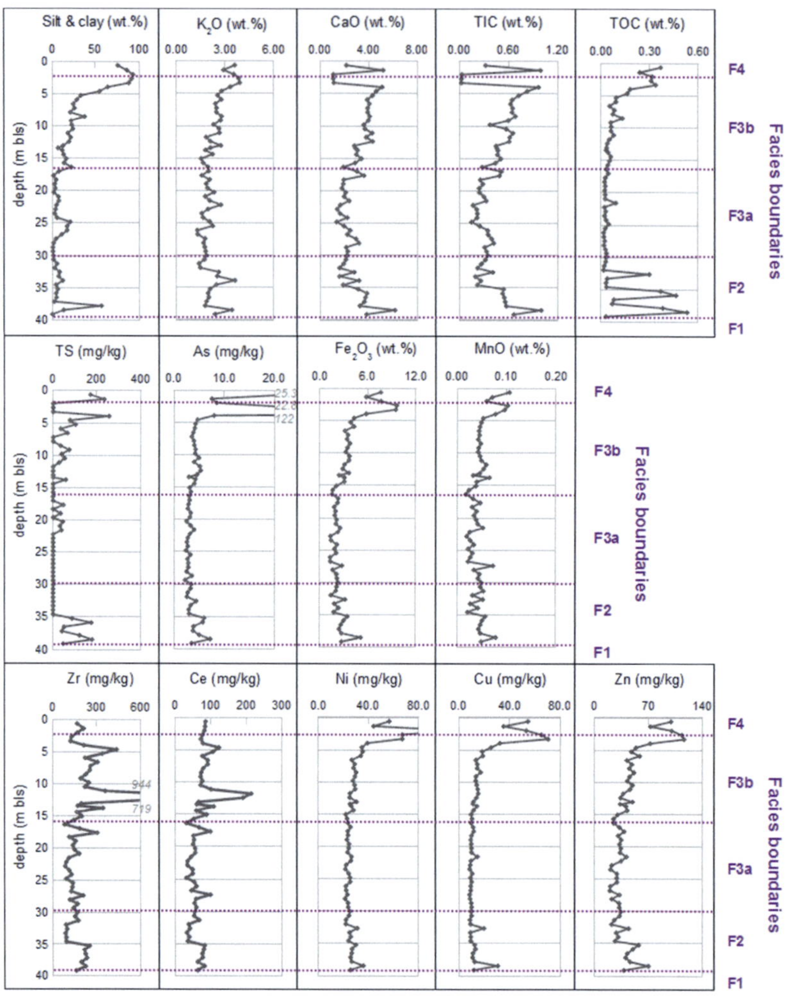

Figure 7.2: Depth profiles of the fine grain fraction (<0.063 mm), major and trace elements, TIC, TOC and TS contents. The profile comprises 61 samples in intervals of 0.65 m.

Mineralogy. The constituent minerals within the aquifer are quartz, followed in frequency by feldspar (according to XRD spectra presumably anorthite, see APPENDIX III, Figure A 3.1), carbonate (calcite and dolomite), mica (muscovite) and chlorite (clinochlore). In clayey samples from F2 and F4, smectite and potentially traces of kaolinite and illite occur, which are characteristic for sediments of the entire Bay of Bengal (DATTA & SUBRAMANIAN 1997) (APPENDIX III, Figure A 3.2). High amounts of dissolved Fe in the Fe-mineral selective steps of the SEP (steps III, IV and V) further indicate the presence of weakly ordered and well crystalline Fe-(oxyhydr)oxides, respectively (see APPENDIX III, Table A 3.2).

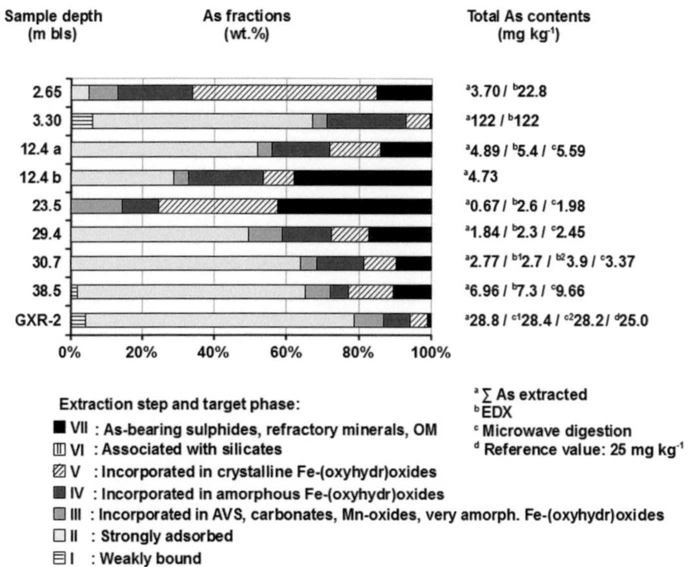

Figure 7.3: SEP results of samples representing different depths and facies. Sedimentary As primarily occurs in the strongly adsorbed fraction (step II) and the fractions that comprise various forms of Fe-(oxyhydr)oxides (steps III, IV and V). Additionally, results for total As are included to approve the recovery rate for As.

Arsenic. In contrast to the aquitard layers of F4, where As contents of up to 122 mg kg^{-1} occur, bulk As contents of the aquifer sediments range from 2.3 to 8.0 mg kg^{-1} (average: 3.8 ± 1.2 mg kg^{-1}; n: 56). Here, As contents correlate positively with Fe, Cu and Zn (Table 7.1). A cluster analysis in form of a dendrogram provides a compact summary of the statistical dependencies of different variables (Figure 7.4). The dendrogram displays close statistical distances of As to Fe, Mn and affiliated trace elements (Cu, Ni and Zn), as well as to TOC and the fine grain fraction (silt and clay). Results of the SEP (Figure 7.3) demonstrate that the bigger part (average: 42.9 %) of As is PO_4^{3-}-extractable (fraction II, interpreted as strongly adsorbed fraction), but this content is most likely overestimated at the expense of fraction III (see chapter 4.2).

Remaining detectable As is associated with (a) acid volatile sulphides, carbonates, Mn-oxides and very amorphous Fe-(oxyhydr)oxides (fraction IIII, average: 7.26 %), (b) amorphous Fe-(oxyhydr)oxides (fraction IV, average: 12.9 %); (c) crystalline Fe-(oxyhydr)oxides (fraction V, average: 14.5 %) and (d) As-sulphides, OM and refractory minerals (fraction VII, average: 22.1 %). In all samples, Mn-oxides play a minor role compared to Fe-(oxyhydr)oxides, as indicated by the absolute contents in the respective fraction (APPENDIX III, Table A 3.2) In case of the clayey layer, the sample from 3.30 m bls has a similar distribution pattern, although the absolute As content is pronounced higher. In the sample from 2.65 m bls, As is increased in the Fe-(oxyhydr)oxide fractions, but has a significant lack of As in the strongly adsorbed phase. The total As content of the seven fractions in this sample is pronounced lower than the bulk As content detected by EDX. This value is therefore considered as an outlier that originates from the inhomogeneous nature of the sample material. Only two samples hold some weakly bound As, while no As is detectable in fraction VI that comprises silicates.

Table 7.1: Summary of a selection of element contents in the aquifer sediments (ranging from 3.35 to 39.1 m bls, n: 56). Included are respective median, lower and upper quartiles (25 % and 75 % Q.), minimum and maximum contents as well as average contents (arithmetic mean) and As correlation coefficients.

Value range	Silt & clay	K_2O	CaO	TIC	TOC	Fe_2O_3	MnO	TS	Ni	Cu	Zn	As
		(wt.%)	(wt.%)	(wt.%)	(wt.%)	(wt.%)	(wt.%)	(mg/kg)	(mg/kg)	(mg/kg)	(mg/kg)	(mg/kg)
Minimum	0.32	1.30	1.37	0.15	0.02	1.36	0.02	<26	21.6	8.58	19.3	2.3
25 % Q.	3.81	1.79	2.09	0.28	0.03	2.07	0.03	<26	23.4	9.62	29.3	3.0
Median	7.62	2.06	2.97	0.43	0.04	2.44	0.04	<26	28.3	11.0	34.0	3.5
75 % Q.	17.5	2.50	3.81	0.58	0.07	3.26	0.05	44.6	39.4	14.0	44.9	4.3
Maximum	57.0	3.62	6.16	1.00	0.53	5.83	0.08	256	36.1	32.3	72.2	8.0
Average	11.5	2.16	2.99	0.45	0.08	2.70	0.04	30.6	26.6	11.6	37.3	3.8
r_{As-}	+0.52	+0.52	+0.73	+0.76	+0.72	+0.83	+0.71	+0.79	+0.77	+0.80	+0.85	-

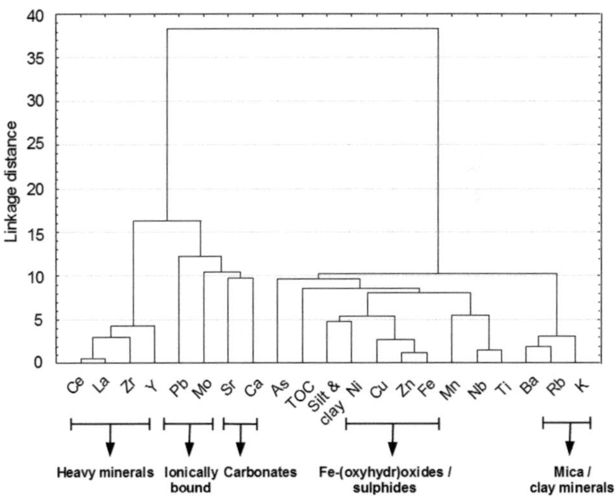

Figure 7.4: Dendrogram for 21 variables (Ward's method, Euclidean distances; n: 61) expressing statistical distances between parameters in aquifer and aquitard sediments.

151

Sedimentary organic matter. In samples from the organic-rich surface near clayey and silty layers of facies F4, $\delta^{13}C$ and $\delta^{15}N$ values, TOC and TN contents are higher than in F3b below (Figure 7.5). In the sandy aquifer sediments of F3b and F3a, TOC and TN concentrations and isotopic values remain comparatively stable. Only two few samples were analysed from F3a due to the throughout low TOC contents. Within facies F2, the lithology changes and TOC and TN contents increase again, whereas $\delta^{13}C$ and $\delta^{15}N$ values further decrease. Since no peat was observed during drilling and sampling, OM exclusively appears in dispersed form.

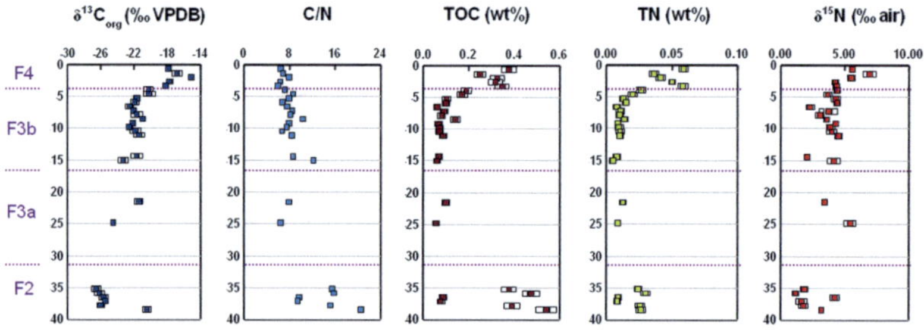

Figure 7.5: Detailed characterisation of OM (n: 27).

Average C/N ratios of samples from facies F4, F3b and F3a are close around 8, which points at freshwater and marine algae as well as C_4 plants as potential origins (Figure 7.6 and APPENDIX II, Table A 2.3). Youngest sediments from F4 carry mixed $\delta^{13}C$ values, reflecting an upward shift to more C_4 influenced sources. Organic matter in F3b and F3a is obviously derived from marine POM. In facies F2, OM consists of more freshwater influenced POM, although the C/N ratio was altered in four samples. In addition, $\delta^{15}N$ values in F4 sediments indicate a potentially increasing influence of terrestrial C_4 plants. $\delta^{15}N$ values from F3b and F3a (average: 3.82 ± 0.84 ‰; n: 16) are in a range that is characteristic for marine POM originating from sediments of the Bay of Bengal (+3.43 to +4.29 ‰ according to GAYE-HAAKE et al. 2005). Four out of six samples from F2 appear to be depleted in $\delta^{15}N$ (values ranging from +1.16 to +4.20 ‰, average: 2.29 ‰; n: 6), which is likely the result of diagenetic effects. The remaining two samples that are considered unaltered are corresponding to F3b and F3a.

In general, $\delta^{13}C$ and C/N values in OM deposited in estuarine sediments are considered to remain stable after sedimentation (LAMB et al. 2006). Nevertheless, GAYE-HAAKE et al. (2005) reported that OM is sensitive to selective degradation during early diagenesis, which may cause a loss of labile compounds (e.g., sugars, amino acids) and an enrichment of stable hydrocarbons (e.g., lignin, cellulose) in vascular vegetation. This causes slight shifts in $\delta^{13}C$, but changes are usually insufficient to prevent determination of the organic source (LAMB et al. 2006). In contrast, small changes in TOC or N contents have the potential to strongly alter the C/N ratios, while fractionation processes may also influence $\delta^{15}N$ values (GAYE-HAAKE et al. 2005, MARCHAND et al. 2005). After early diagenetic modifications, C/N- and $\delta^{13}C$ values of OM are retained for million-year time periods (MEYERS 1994, SARKA et al. 2009).

Total organic carbon correlates well with TN (r: +0.75) as well as $\delta^{13}C$ with $\delta^{15}N$ values (r: +0.68), indicating that TN is primary organically derived (total N $\approx$ organic N) and that the original source signatures remained preserved. Only four samples from F2 constitute an exception (Figure 7.5 and APPENDIX III, Figure A 3.3). These particular samples are depleted in N relative to TOC and in $\delta^{15}N$, indicating a fractioning loss of N. Contents of TOC and TN in the sand layers are generally similar to average values in present floodplains of Bangladesh (TOC: 0.05-0.63 wt.%, TN: 0.01-0.06 wt.%; C/N: 5-11.4; DATTA et al. 1999), which indicates that only minor diagenetic alteration occurred in the remaining 23 samples. Results further coincide with the results from the JAM study site, where $\delta^{13}C$ values decrease between 1.5 to 45.7 m bls from -22 ‰ to -28.5 ‰, and TOC from 7.2 to 0.5 wt.% (MC ARTHUR et al. 2004, SARKA et al. 2009).

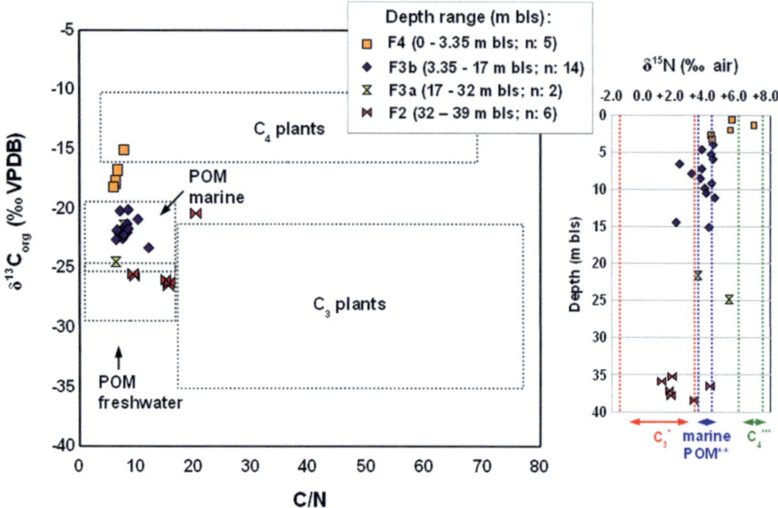

Figure 7.6: Comparison of measured and literature values in form of a $\delta^{13}C$ - C/N plot and a $\delta^{15}N$ – depth plot to distinguish different sources of OM (n: 27).

*$\delta^{15}N$ range for mangrove (C$_3$ type) tissues (MUZUKA & SHUNULA 2006)

**Sedimentary POM from Bay of Bengal (GAYE-HAAKE et al. 2005)

***C$_4$ plant pollen (DESCOLAS-GROS & SCHÖLZEL 2007)

7.2.2 *GROUNDWATER CHARACTERISATION*

7.2.2.1 Groundwater properties

Hydrochemical baseline. Table 7.2 provides an overview of physico-chemical parameters of groundwater sampled in December in 2009, two months after the monsoon season and right before the abstraction experiment was conducted. Except for one sample, ion balances remain below 7 %, which is an acceptable range. The results are considered as representative baseline values for the five monitoring wells. In order to identify outliers and confirm anomalies, results were compared to the regularly monitoring sampling.

In December 2009, the hydrostatic head inside the wells was met at 2.13 m bls. Water temperatures ranged from 26.0 to 27.2 °C and pH values were circum-neutral. The throughout anoxic groundwater belongs to the Ca-Mg-HCO$_3$-type and is characterised by a high alkalinity (Figure 7.7). Within groundwater from all five wells, noticeable concentrations of Fe, Mn, PO$_4^{3-}$ and As occur, while NO$_3^-$ concentrations are close to or below the limit of quantification (Table 7.2). Computed saturation indices (SI) indicate that groundwater is supersaturated regarding magnetite, hematite, siderite and slightly supersaturated regarding calcite and ordered dolomite, while disordered dolomite is partly under saturated (Table 7.3). In addition, magnetite and hematite are strongly supersaturated. Calculated aqueous to solid phase mol ratios revealed that As is much more enriched in ground-water than any other element (Table 7.2 and APPENDIX III, Table A 3.4).

Table 7.2: Selection of hydrochemical baseline values that are used in the following to interpret induced changes in the groundwater composition (complete results presented in APPENDIX III, Table A 3.3). Samples were taken immediately before the pumping experiment was conducted (on 03/12/09). *Thermodynamically dominating species* determined with PHREEQC.

Well (m bls)	pH	EC (μS/cm)	T (°C)	TA HCO_3^- (mg/L)	Na Na^+ (mg/L)	K K^+ (mg/L)	Ca Ca^{2+} (mg/L)	Mg Mg^{2+} (mg/L)	Cl^- (mg/L)	SiO_2 (mg/L)	δ^2H (‰ VSMOW)	$\delta^{18}O$ (‰ VSMOW)
A (12-21)	6.9	973	26.0	586	29.4	9.73	134	28.4	49.7	12.1	-23.3	-2.95
B (22-25)	6.9	1040	26.6	610	27.8	9.70	140	27.7	66.1	11.2	-23.6	-3.07
C (26-29)	7.2	718	27.2	482	18.3	2.73	93.9	23.3	18.6	12.2	-28.3	-3.92
D (30-33)	7.1	781	26.9	500	19.9	2.72	109	24.7	19.5	13.6	-30.9	-4.17
E (34-37)	7.0	745	26.9	513	14.1	2.88	92	22.7	20.3	13.1	-27.7	-3.81
Aqueous/ solid phase ratios*				nd	1.60 to 2.96 x 10^{-3}	1.09 to 7.01 x10^{-4}	0.23 to 1.42 x 10^{-2}	1.34 to 7.59 x 10^{-3}	nd	nd		

Well (m bls)	DOC (mg/L)	NO_3^- (mg/L)	NH_4^+ (mg/L)	Mn Mn^{2+} (mg/L)	Fe Fe^{2+} (mg/L)	SO_4^{2-} (mg/L)	PO_4^{3-} $H_2PO_4^-$ (mg/L)	Ba Ba^{2+} (mg/L)	As** H_3AsO_3 (μg/L)	Mo MoO_4^{2-} (μg/L)	U $UO_2(HPO_4)_2^{2-}$ (μg/L)	Ion balance (%)
A (12-21)	3.02	<0.88	0.65	0.92	7.77	7.78	2.53	0.33	98.0	1.21	0.08	-1.63
B (22-25)	4.52	<0.88	1.06	0.74	8.96	10.7	2.90	0.43	100	0.86	0.08	-4.76
C (26-29)	2.55	<0.88	0.99	0.85	1.09	<0.85	2.45	0.17	296	1.97	0.58	-6.32
D (30-33)	1.53	0.97	1.18	0.37	4.76	<0.85	2.98	0.26	262	1.23	0.16	-1.59
E (34-37)	1.32	<0.88	3.41	0.47	4.80	<0.85	2.76	0.23	158	1.57	0.16	-15.3
Aqueous/ solid phase ratios*	0.03 to 2.17 x 10^{-2}	nd	nd	0.69 to 2.90 x 10^{-3}	0.73 to 8.67 x 10^{-4}	nd	1.33 to 4.24 x 10^{-3}	0.41 to 1.42 x 10^{-3}	0.16 to 1.21 x 10^{-1}	<dl	0.15 to 1.81 x 10^{-4}	

*Solid phase contents determined by acid microwave digestion

**Average As(III) percentage: 94.7 %

nd: not determined

Table 7.3: SI for potentially present mineral phases, calculated from data of samples taken on 03/12/09 (assumptions: O_2 of 0.24 mg/L $^{-1}$, E_H of -59 mV).

Well (m bls)	Quartz	Anorthite	Calcite	Dolomite ord. / disord.		Chlorite	Phlogopite
A (12-21)	0.27	-3.92	0.27	0.23	-0.31	-8.03	-10.2
B (22-25)	0.23	-3.07	0.30	0.27	-0.28	-7.27	-9.92
C (26-29)	0.26	-4.41	0.42	0.61	0.07	-5.80	-9.06
D (30-33)	0.31	-3.69	0.35	0.44	-0.11	-6.10	-9.37
E (34-37)	0.29	-8.32	0.20	0.15	-0.40	-11.7	-12.4

Well (m bls)	$Fe(OH)_3$ amorph.	Magnetite	Hematite	Siderite	Anhydrite	Halite
A (12-21)	-2.44	10.8	8.99	1.12	-2.78	-7.44
B (22-25)	-2.40	11.0	9.10	1.18	-2.63	-7.34
C (26-29)	-2.24	11.2	9.47	0.57	-4.23	-8.06
D (30-33)	-2.00	12.0	9.93	1.08	-4.05	-8.01
E (34-37)	-2.30	11.2	9.34	1.00	-4.33	-8.14

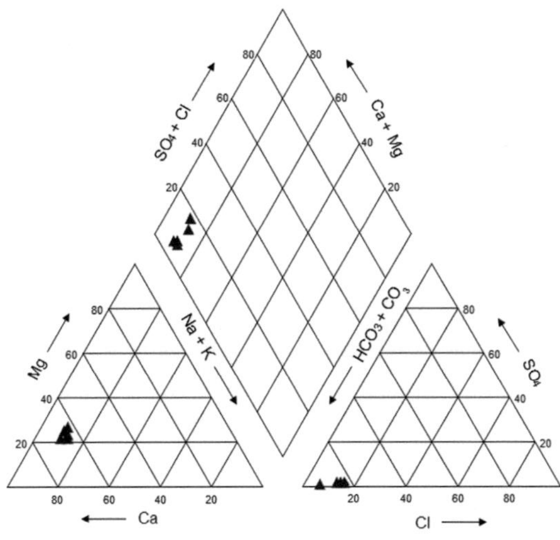

Figure 7.7: Piper diagram displaying major solutes in groundwater of the five monitoring wells (date: 03/12/09).

Hydrochemical stratification within the aquifer. The distribution of major solutes as well as stable isotopic values of groundwater samples taken in December 2009 reflect the presence of two distinctive, hydro-chemical layers within the groundwater body. These layers are divided by a presumed boundary located in between the well screen positions of the two monitoring wells B and C in about 25 to 26 m bls. Corresponding to higher concentrations of major solutes (Ca, Mg, Na, K, HCO_3^-, SO_4^{2-}, and Cl^-), the EC is in wells A and B pronounced higher than in the deeper monitoring wells. In addition, groundwater from this upper layer (entitled as layer I) includes higher $\delta^{18}O$ and δ^2H values than water from the layer below (entitled as layer II) (Figure 7.8). Stable isotopic values in all water samples plot below the LMWL, with a regression line that intersects the LMWL closely to the annual volume-weighted rainfall average. In addition, depth distributions of trace elements such as Fe, Mn, Ba, Sr, U, PO_4^{3-}, and As superimpose these two layers (Table 7.2).

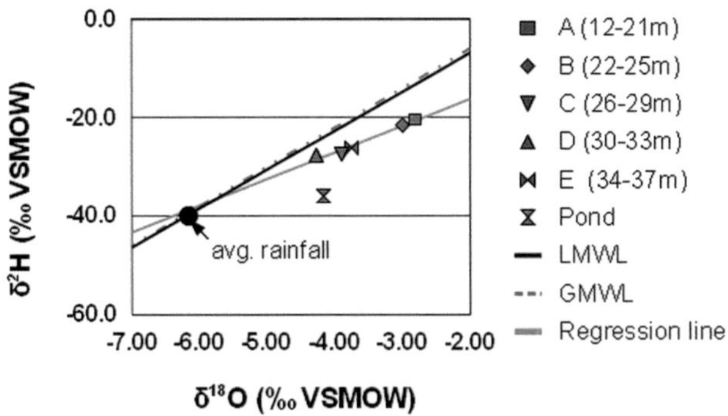

Figure 7.8: Isotopic compositions of O and H in samples taken on 03/12/09 in comparison to the global meteoric water line (GMWL), the local meteoric water line (LMWL), and the volume-weighted average composition of rainfall 2004/05 (SENGUPTA et al. 2008) for an area situated about 40 km south of the study site (JAM area).

Maximum As concentrations occur in groundwater from well C, where the groundwater chemistry significantly differs from the other four wells. Here, lowest concentrations of dissolved Fe meet highest concentrations of Mo, U and As. From this depth on, SO_4^{2-} concentrations were close to the dl. Groundwater from the two wells D and E beneath is characterised by similar compositions regarding the major solutes.

Comparison with adjacent wells, pond and rain water. Five tube wells with comparable depths (between 20 and 55 m) were sampled within 500 m around the high As site (Table 7.4). Samples from four of these wells (including well 132, which was the reason to choose this area as high As site) are characterised by highly similar hydrochemical compositions. In contrast to the shallow monitoring wells A and B, none of the tube wells was found to deliver water from the upper layer I, where the salinity is increased. Hence, these two monitoring wells have to be considered with caution.

To assess the possibility of surface recharge from the small pond located next to the monitoring wells, surface water was sampled, too (see Table 7.4). Stable isotopic values of the pond water are included in Figure 7.8. Surprisingly, pond water contains 36.3 µg L^{-1} As, with an As(III) percentage of 27.4 %.

Table A 7.4: Comparison of relevant parameters in the monitoring wells (on 03/12/09) with nearby wells that were sampled in 2007. Wells 132 formed the base of decision to choose the respective area as study site.

Well (m bls)	Year	Na	K	Ca	Mg	Cl^-	NO_3^-	SO_4^{2-}	PO_4^{3-}	Fe	Mn	As	As/PO_4^{3-}	Na/Cl^-
					all (mg/L)							(µg/L)	(mol ratio)	
A (12-21)	2009	29.4	9.73	134	28.4	49.7	<0.88	7.78	2.53	7.77	0.92	98.0	0.05	0.91
B (22-25)	2009	27.8	9.70	140	27.7	66.1	<0.88	10.7	2.90	8.96	0.74	100	0.04	0.65
C (26-29)	2009	18.3	2.73	93.9	23.3	18.6	<0.88	<0.85	2.45	1.09	0.85	296	0.15	1.52
D (30-33)	2009	19.9	2.72	109	24.7	19.5	0.97	<0.85	2.98	4.76	0.37	262	0.11	1.57
E (34-37)	2009	14.1	2.88	92	22.7	20.3	<0.88	<0.85	2.76	4.80	0.47	158	0.07	1.08
Pond	2009	12.4	15.8	30.7	8.52	25.0	<0.88	1.89	1.11	0.22	0.20	36.3	0.04	0.77
132 (24)	2007	26.8	3.33	91.7	22.5	18.5	1.40	1.44	4.23	3.30	0.49	285	0.09	2.24
131 (24)	2007	18.5	1.78	77.0	17.5	10.9	<0.88	<0.85	3.74	5.46	0.51	84.9	0.03	2.62
179 (55)	2007	17.0	2.32	66.9	19.0	6.67	<0.88	3.99	0.70	0.03	0.16	120	0.22	3.92
180 (24)	2007	25.8	1.63	74.3	19.8	7.04	1.60	<0.85	4.90	3.88	0.11	116	0.03	5.66
181 (20)	2007	14.0	3.03	71.6	17.3	4.23	1.34	<0.85	4.77	3.65	0.21	124	0.03	5.11
Rain water	2009	1.97	1.26	1.85	0.18	7.35	<0.88	2.62	0.03	0.01	0.06	0.16	0.01	0.41

7.2.2.2 The groundwater abstraction experiment

Hydrological characterisation of the aquifer. During each of the four pumping cycles, a rapid decline of the hydrostatic head was recorded in adjacent observation wells B to E, which was documented in detail for the first pumping cycle (Table 7.5). After five hours of continuously groundwater abstraction, declines in the respective hydrostatic heads had stopped, respectively increased by several centimetres. This indicated that the aquifer was in a confined condition.

After 48 h of continuous pumping (end of the third pumping cycle), declines in hydrostatic heads were considered as nearly constant and resulted in a small depression cone (illustrated in Figure 7.9). Hydrostatic heads remained throughout above the base of the surface aquitard located in about 3.35 m bls (lowest observed level: 2.92 m below top edge of tube, which is equal to 2.63 m bls). Hence, the virtue water table remained constantly uninfluenced. Based on the draw-downs presented in Figure 7.9, a K value of ~1.3 x 10^{-4} m sec^{-1} was estimated using the Dupuit-Thieme equation for confined aquifers (ENTENMANN 2006) (see APPENDIX III, Figure A 3.4). Some necessary assumptions for the use of this equation were not exactly met (the aquifer is inhomogeneous, screening lengths do not penetrate the complete aquifer and observation wells are too close), but this estimated K value is still in good agreement with the grain size based K estimations (included in Figure 7.9) and results of a pumping test conducted by NEUMANN et al. (2009) in Bangladesh, where the horizontal conductivity K_h was estimated as 3.3 x 10^{-4} m sec^{-1} and the vertical conductivity K_v as 1.3 x 10^{-5} m sec^{-1} (for a depth of 15 to 120 m).

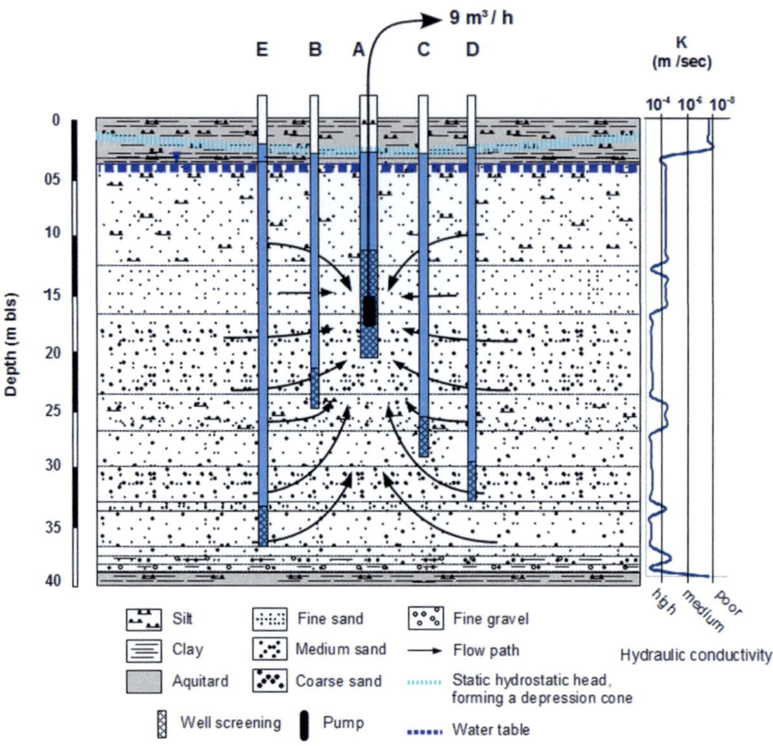

Figure 7.9: Schematic sketch of the abstraction experiment setup including subsequent declines of the hydraulic heads of the observation wells during the first pumping cycle (t_0: 10^{15}, 05/12/09).

Table 7.5, a: Hydrostatic heads during the abstraction experiment, recorded hourly until 17^{15}, and again after 29 h (15^{15}, 06/12/09). Decreases are given in centimetres relative to t_0 before the pumping was started. No measurements could be done in the central well A while pumping was in progress. Table 7.5, b: Hydrostatic heads after 48 h of continuously pumping at the end of the third pumping cycle (13/12/09). Hydrostatic head in well A was measured immediately after pumping was stopped.

a)

Time of pumping (h)	Well (well screen m bls)			
	B (22-25)	C (26-29)	D (30-33)	E (34-37)
0	0	0	0	0
1	16	9	7	7
2	24	10	7	9
3	25	16	13	13
4	23	15	15	14
5	25	17	17	17
6	20	13	10	10
7	18	10	7	8
29	20	12	10	11

b)

Well (m bls)	A (12-21)	B (22-25)	C (26-29)	D (30-33)	E (34-37)
Distance to well A (m)	-	1.96	4.70	4.85	2.00
Baseline static head (03/12/09) (m below top edge tube)	2.42	2.41	2.42	2.43	2.44
Static head after 48 h pumping (13/12/09) (m below top edge tube)	2.92	2.69	2.63	2.63	2.65
Net drawdown (m)	0.50	0.28	0.21	0.20	0.29

Changes in groundwater chemistry. Accompanying sampling during the abstraction experiment showed remarkable changes in the hydro-chemistry of the groundwater, especially in case of the upper two wells A and B (Figure 7.10). Parameters that had initially differed developed partly pronounced decreases (e.g., Ca, Cl$^-$, δ^{18}O), respectively increases (As, Si). Importantly, most non-conservative elements like Ca and As behaved in this case similar to typically conservative tracers like Cl$^-$ and δ^{18}O (SENGUPTA & SARKA 2006).

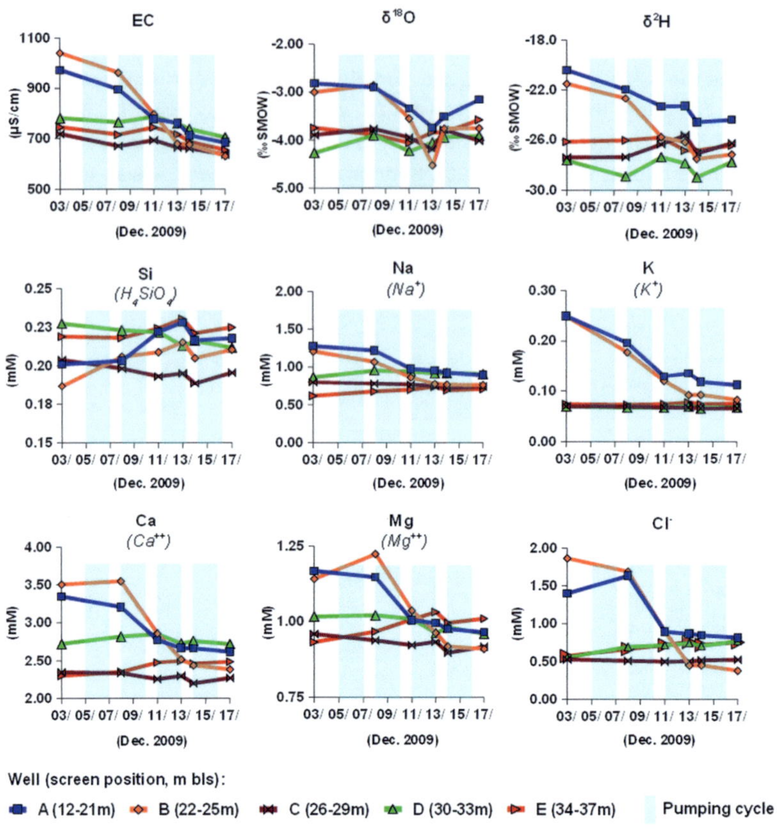

Figure 7.10, part I: Changes in the hydrochemistry of the monitoring wells during the abstraction experiment. Stated are thermodynamically prevailing species of the dissolved major and trace elements. All parameters show that wells A and B are initially located in a different hydrochemical layer compared to wells C, D and E. During pumping, the groundwater composition of the upper two wells adjusts toward the hydrochemistry of the deeper wells D and E. In brackets. *thermodynamically prevailing species* of dissolved major and trace elements are stated. Figure continued on next page.

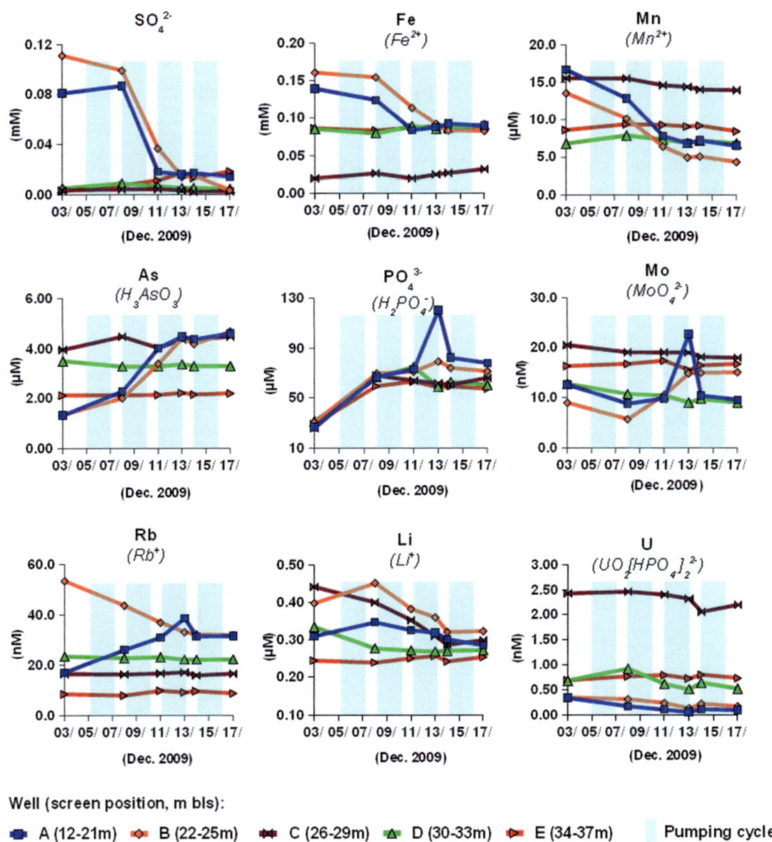

Figure 7.10, part II: Included are trace elements that are relevant for As mobility. In brackets: *thermodynamically prevailing species* of dissolved major and trace elements are stated.

Assuming a conservative-like behaviour for K, respective concentration changes can be used to estimate the increasing ratio of water originating from layer II in range of layer I (Figure 7.11). The baseline concentrations of K in wells A and B (layer I) were 9.73 and 9.70 mg L^{-1}, while initial concentrations in the other three wells (layer II) were 2.73 (well C), 2.72 (well D) and 2.88 mg L^{-1} (well E), respectively. Following a continuously increase, more than 90 % of the groundwater met in well B originated from layer II at the end of the experiment, and approximately 75 % in well A (Figure 7.11).

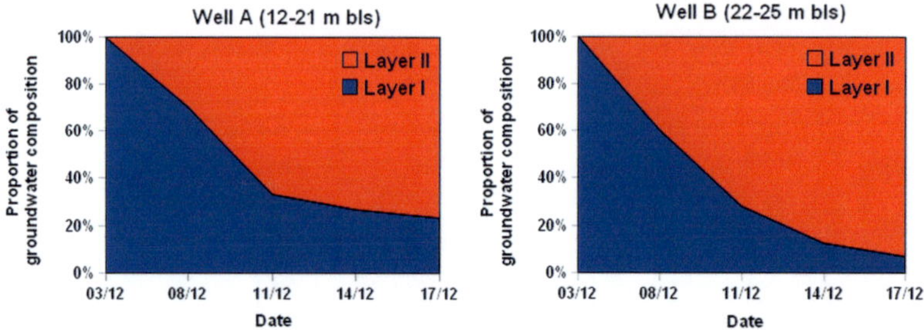

Figure 7.11: Increasing shares of groundwater originating from layer II in wells A and B. Calculations based on relative changes in concentrations of K.

For further differentiations, compounds with initially deviant distribution patterns are useful. For example, Rb concentrations rapidly increased in well A and reflected thereby an influence of groundwater from the range of well B. In addition, major ions like Ca and Na reflected a subsequent increasing share of water from range of well D and E (Figure 7.10 and APPENDIX III, Table A 3.3). Increasing As concentrations demonstrated a strong potential influence of water from range of well C, where the maximum concentrations were initially located, assuming a conservative-like behaviour of As for this time. After the last pumping cycle, As concentrations in wells A, B and C exceeded the baseline values. In addition, As(III) was constantly the dominating As species (average percentage: 96.2 ± 3.9 %, n: 29).

The sample that was taken immediately after 48 h of continuously pumping (13/12/09) from well A further revealed that PO_4^{3-} and Mo concentrations had much more increased than it was visible after the interception of 24 h (14/12/09). Despite this, only marginal changes occurred in well C, especially regarding trace elements with comparatively low (e.g., Fe), or high concentrations (Mn, Li, and U). The only parameters that reacted to pumping were Li and PO_4^{3-}, and to a lesser extent As. The sample that was taken immediately after 48 h of pumping displayed in addition a significant shift in the $\delta^{18}O$ and δ^2H values.

In the deepest wells D and E, most major and trace element concentrations remained nearly unaffected, except for PO_4^{3-}. Phosphate concentrations had simultaneously increased in all five wells immediately after the first pumping interval was completed. At the end of the experiment (after four pumping and rest cycles), PO_4^{3-} concentrations had increased by up to 192 % in comparison to the baseline values. The same applied to DOC concentrations, which increased by up to 359 % (Table 7.6).

Pumping was further accompanied by a change in the odour of the extracted groundwater, which changed from odourless towards sulphidic, with an intensity ranging from weak (wells C and D), over medium (A and E) to strong (B). This experiment impressively demonstrated how groundwater extraction can affect the groundwater composition including the As concentrations in local shallow tube wells.

Table 7.6: Parameters characterised by net increases after the fourth and last pumping cycle.

Well (m bls)	Date	PO_4^{3-} (μM)	DOC (mM)	As (μM)	Odour
A	03/12/09	26.7	0.25	1.31	without
(12-21)	17/12/09	77.9	0.69	4.58	medium (H_2S)
	net change:	+192 %	+175 %	+250 %	
B	03/12/09	30.6	0.38	1.34	without
(22-25)	17/12/09	71.3	0.56	4.66	strong (H_2S)
	net change (%)	+133 %	+47.7 %	+249 %	
C	03/12/09	25.8	0.21	3.95	without
(26-29)	17/12/09	66.1	0.47	4.46	weak (H_2S)
	net change (%)	+156 %	+120 %	+13.0 %	
D	03/12/09	31.4	0.13	3.49	without
(30-33)	17/12/09	60.6	0.51	3.29	weak (H_2S)
	net change (%)	+93.0 %	+303 %	-5.98 %	
E	03/12/09	29.0	0.11	2.11	without
(34-37)	17/12/09	57.3	0.50	2.18	medium (H_2S)
	net change (%)	+97.3 %	+359 %	+3.29 %	

7.2.2.3 Monitoring results

In the following, results from the hydrochemical monitoring (December 2008 to August 2010) are presented in context of temporary varying As concentrations and mid-term consequences related to the abstraction experiment (Figure 7.12). Results of the first year of monitoring were described by BISWAS et al. (2011).

Time resolved variations in the hydrochemistry. In dependence of the prevailing climatic conditions, the hydrostatic head recorded inside monitoring well B showed pronounced annual oscillations that ranged from 0.74 m bls in December 2009 (post-monsoon season) to 5.23 m bls in April 2010 (end of the dry pre-monsoon season) (Figure 7.12). As a result, conditions in the aquifer seasonally fluctuated between confined and unconfined. This was accompanied by changes in the hydrochemical composition of groundwater that occurred in monitoring samples of wells A, B and C. In the following, results from the dry pre-monsoon and rainy monsoon seasons are presented separately.

Pre-monsoon seasons. The upper, saline layer I (indicated by a higher salinity and $\delta^{18}O$ values) as well as the vertical stratification of redox sensitive elements (Fe, As and SO_4^{2-}) were stable throughout the monitoring, except for both pre-monsoons seasons. To this time, highly similar trends arose in the hydrochemistry of the monitoring samples in 2009 as well as after the abstraction experiment in 2010. Each time, compositions in wells A and B subsequently changed towards the typically saline, evaporation influenced water with lower As concentrations that is characteristic for layer I. During the dry seasons, hydraulic heads continuously declined until reaching low stands end of April.

Hydraulic conditions changed from confined to unconfined between January and February, as soon as the hydrostatic head and therefore the water table fell below the aquitard layers reaching down to 3.35 m bls. Despite the pronounced changes caused by the abstraction experiment, groundwater in well A rapidly returned to the initial baseline after two weeks in January 2010, as indicated by declining As concentrations, increasing $\delta^{18}O$ values and rising K, Cl⁻, SO_4^{2-} and Fe concentrations (Figure 7.12). With a lag of 4-6 weeks, similar trends formed in well B between March and May. Here, trends in major ions were negatively correlated to the development of the hydrostatic head, too. The shape of these trends was further mirror-inverted compared to those that had developed during the pumping experiment. When the baseline values were met end of April 2010, the unconfined groundwater table was met in 5.25 m bls. In 2009, the same pronounced trends appeared, especially in well B.

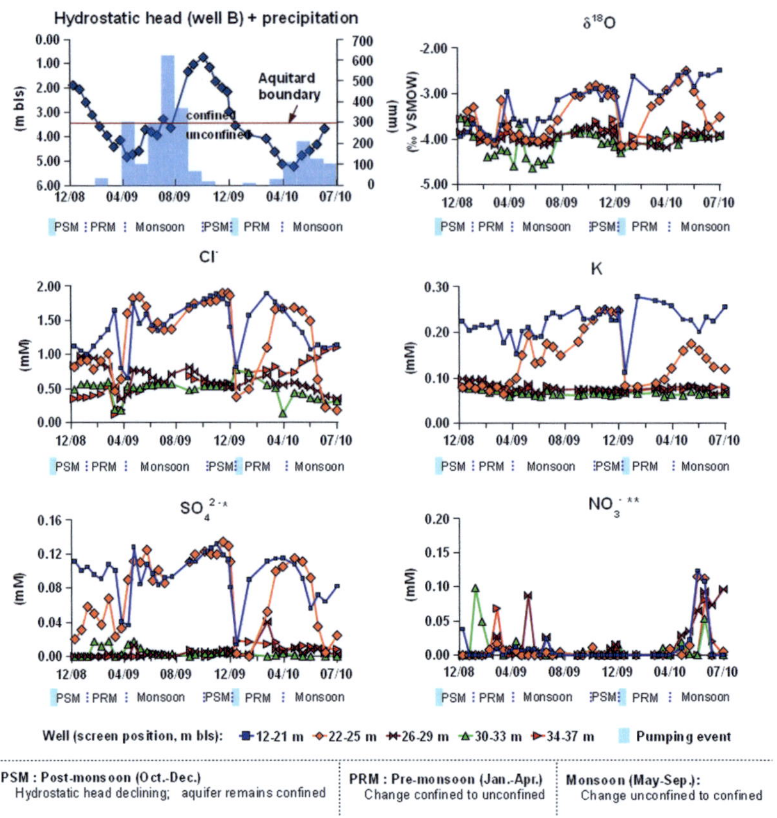

Figure 7.12, part I: Summary of hydrochemical parameters that display characteristic vertical differences concerning $\delta^{18}O$, Cl^-, K, SO_4^{2-} and NO_3^- concentrations. In addition, temporal trends occurred in wells A, B and C in dependence of seasonal variations of the hydrostatic head. Monthly average precipitation for the Nadia district provided by the India Meteorological Service. Figure continued on next page.

*lq for SO_4^{2-}: 0.085 mM
**lq NO_3^-: 0.014 mM

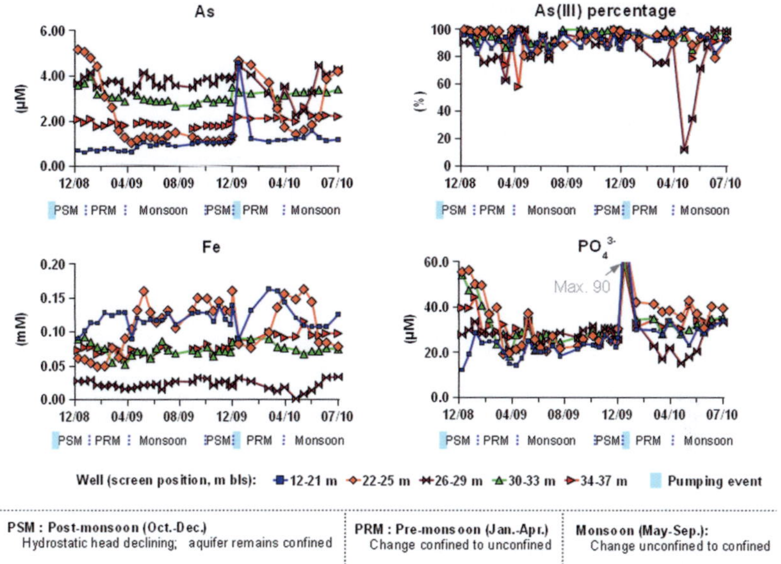

Figure 7.12, part II: Summary of hydrochemical parameters relevant for As mobility that display characteristic vertical differences concerning As, As(III), PO_4^{3-}, and Fe. In addition, temporal trends in wells A, B and C manifested in dependence of seasonal variations of the hydrostatic head.

Monsoon seasons. With onset of the monsoon season in 2009, temporarily slight declines in $\delta^{18}O$ values, K, Cl⁻ and SO_4^{2-} concentrations occurred in wells A and B, before values returned to the initial situation in May. Despite this, pronounced trends occurred in the upper wells soon after the monsoon rains began in 2010 (Figure 7.12). Most obvious changes manifested again in water from well B, where As concentrations rose while $\delta^{18}O$ and major element concentrations decreased at the same time. These trends formed between June and end of July alongside the increasing hydrostatic head, which rose by 1.55 m to that time. Finally, all characteristic hydrochemical parameters in groundwater of well B had gradually adapted to values comparable to those of the layer II.

In groundwater of monitoring well A, Cl^-, K, SO_4^{2-} and Fe concentrations first declined until June 2010, before values stabilised or increased again. Meanwhile, compositions in wells C, D and E had remained predominantly constant, except for Cl^-.

Concentrations of NO_3^- surprisingly increased short termed in groundwater of all wells and peaked in June 2010 (max. of 7.67 mg L^{-1} in well A), before they fell again below the limit of quantification (except for well C, whereas contents continuously increased until July). A similar situation appeared during the pre-monsoon season 2009, when scattered NO_3^- peaks appeared in some of the groundwater samples.

Average Fe(II) percentages exceeded 90 % in groundwater between 15/03/09 to 29/05/09, except for well C, where a significant lower average ratio of 67 % occurred.

Behaviour of arsenic. Plots of As, Fe and PO_4^{3-} concentrations revealed a close relation between these elements. Distinctive and relatively stable mol ratios adjusted, characteristic for each well and therefore depth range (Figure 7.13).

Arsenic concentrations in wells D and E remained at nearly constant levels compared to the hydrochemical baseline (03/12/09). In contrast, As concentrations showed temporarily deviant concentrations in the three other wells A, B and C:

- In well B, eye-catching trends in As concentrations arose during both pre-monsoon seasons and the monsoon season 2010 (Figure 7.12). The highest absolute amount of As at the high As site (387 µg L^{-1}) was determined from the first sample taken from this well (December 2008). During the following pre-monsoon season, As values gradually declined until a stable level adjusted with an average concentration of 90.5 ± 9.3 µg L^{-1} (n: 16). Values heavily increased again during the abstraction experiment in December 2009, and rapidly declined afterwards during the pre-monsoon season. Surprisingly, concentrations rose again with beginning of the monsoon season in 2010 (up to 315 µg As L^{-1}), which had not occurred in 2009 to the same time, when As concentrations had constantly remained <100 µg L^{-1}. During the complete monitoring period, As concentrations shifted between those of well A (~100 µg L^{-1}) and well C (>300 µg L^{-1}). Furthermore, As concentrations correlated negatively with conservative tracers like $\delta^{18}O$ and Cl^-, but positively with PO_4^{3-} and the hydrostatic head. The trends that formed during the monsoon season 2010 exactly traced that of the pumping experiment and were additionally mirror-inverted compared to the trends of the two pre-monsoon seasons (Figure 7.14). These pronounced concentration changes did not affect the As speciation as indicated by constantly high As(III) percentages of 93.4 ± 5.8 % (n: 36).

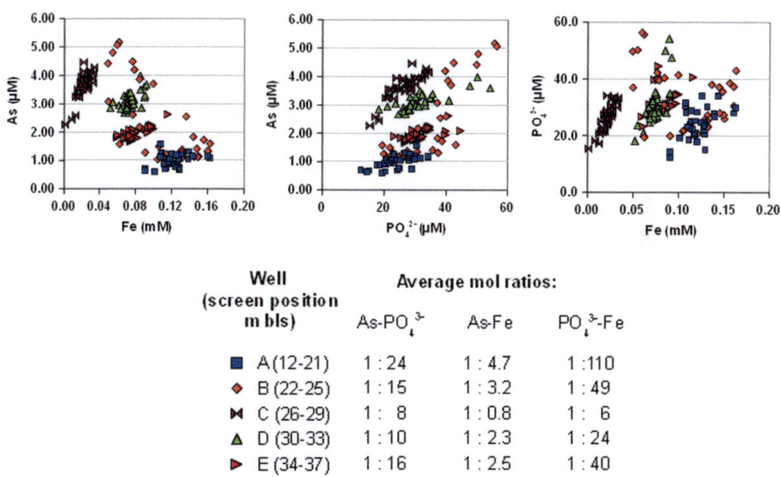

Well (screen position m bls)	Average mol ratios: As-PO_4^{3-}	As-Fe	PO_4^{3-}-Fe
■ A (12-21)	1 : 24	1 : 4.7	1 : 110
◆ B (22-25)	1 : 15	1 : 3.2	1 : 49
⋈ C (26-29)	1 : 8	1 : 0.8	1 : 6
▲ D (30-33)	1 : 10	1 : 2.3	1 : 24
▶ E (34-37)	1 : 16	1 : 2.5	1 : 40

Figure 7.13: Plots of As, Fe and PO_4^{3-} mol ratios for all monitoring samples show close and stable relationships between these three solutes for each of the five wells, except for well B, where respective solutes are influenced by temporary fluctuations.

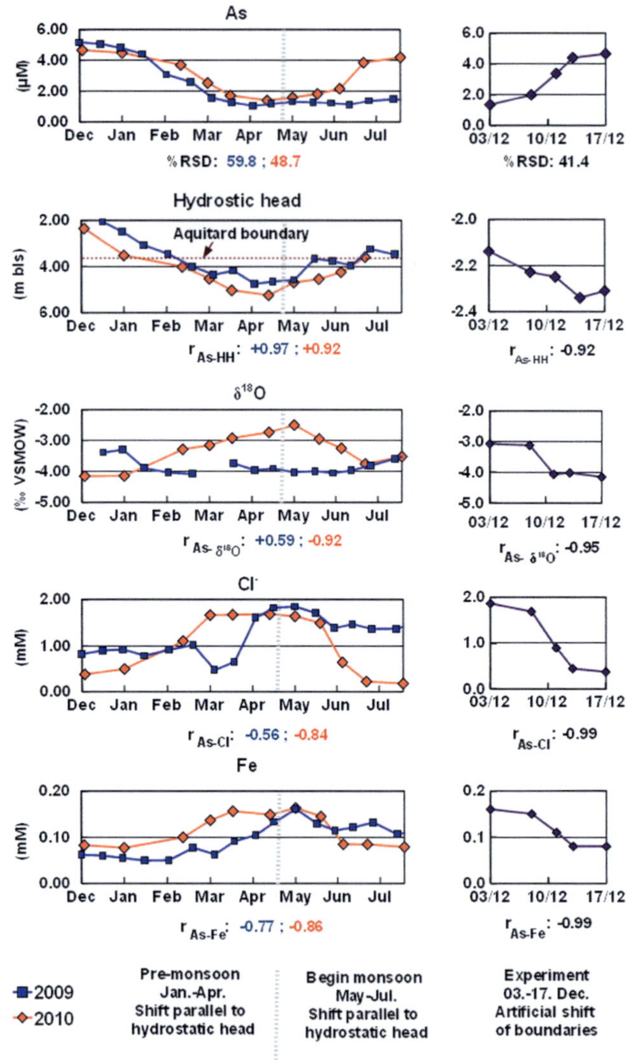

Figure 7.14: Trends in As, indicator parameters and the hydrostatic head of well B during 2009 and 2010 in comparison to the abstraction experiment. Surprisingly, a pronounced shift appeared during the monsoon season in 2010, which was not visible during 2009.

- Despite all other parameters, As concentrations strongly varied in well C during 2010, too (Figure 7.15). Absolute As concentrations halved from 334 µg L^{-1} at the end of pumping experiment to 169 µg L^{-1}, when the lowstand of the hydrostatic head was reached end of April. At the same time, the As(III) percentage strongly declined from 98.0 % down to 12.1 %, which was accompanied by a decrease of Fe concentrations from 1.50 to 0.07 mg L^{-1}. During monsoon (May to July), As concentrations subsequently rose again to 321 µg L^{-1}, which was close to the maximum at the end of the experiment (334 µg L^{-1}). To this time, Fe and As(III) concentrations also climbed back to previous values.

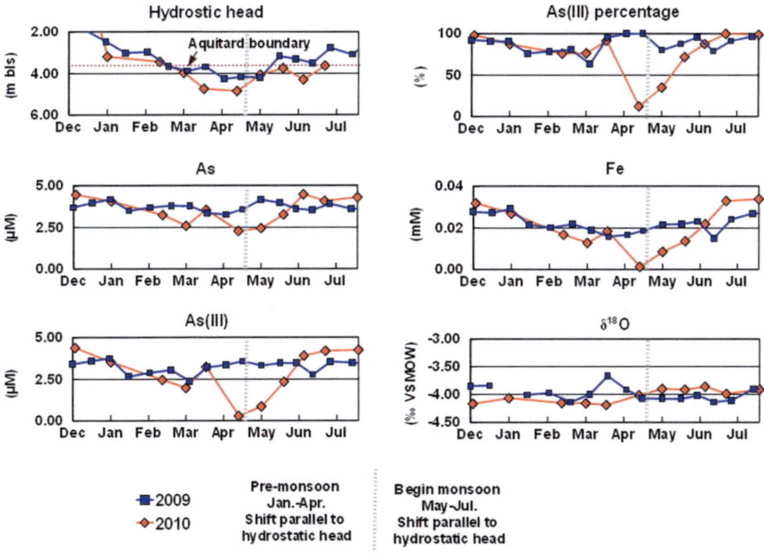

Figure 7.15: Comparison of As trends in well C during 2009 and 2010 together with other relevant parameters.

- During the pre-monsoon season in 2010, As concentrations in groundwater of the central well A rapidly returned to the baseline that was recorded before the experiment. Considering the complete monitoring period, the As level subsequently increased (Dec. 2008: 52.0 µg L^{-1}; Nov. 2009: 82.8 µg L^{-1}; Jul. 2010: 88.3 µg L^{-1}, Figure 7.16). This increase was also apparent in δ^{18}O values and PO$_4^{3-}$, causing near-constant As to PO$_4^{3-}$ mol ratios of between 1:18 and 1:25. The reduced form As(III) permanently remained the prevailing As species.

Spatiotemporal differences of As and other redox sensitive solutes as well as conservative tracers are summarised in Figure 7.17. Here, the attempt is made to interpret temporal changes that occurred between end of April and July 2010 as vertically differing hydrochemical layers.

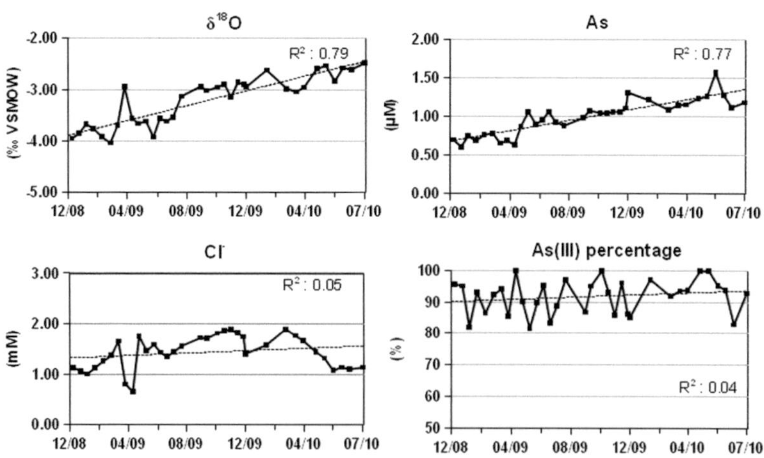

Figure 7.16, part I: Development of As and other important characteristic parameters in well A during the complete monitoring period. R^2: coefficient of determination. Figure continued on next page.

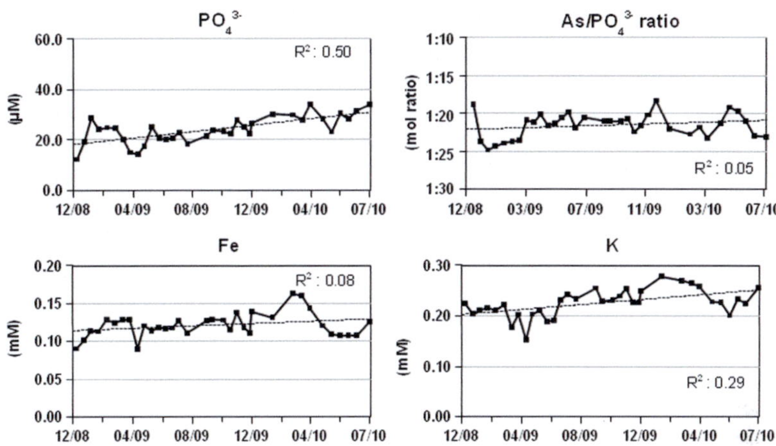

Figure 7.16, part II.

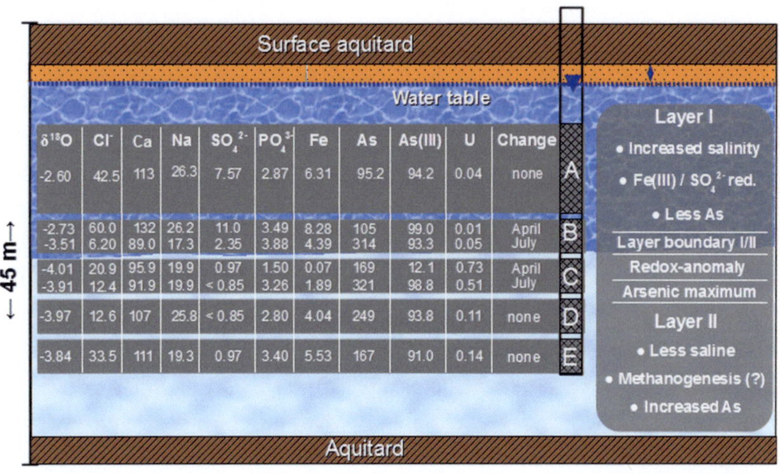

$\delta^{18}O$	Cl$^-$	Ca	Na	SO$_4^{2-}$	PO$_4^{3-}$	Fe	As	As(III)	U	Change	
-2.60	42.5	113	26.3	7.57	2.87	6.31	95.2	94.2	0.04	none	A
-2.73	60.0	132	26.2	11.0	3.49	8.28	105	99.0	0.01	April	B
-3.51	6.20	89.0	17.3	2.35	3.88	4.39	314	93.3	0.05	July	
-4.01	20.9	95.9	19.9	0.97	1.50	0.07	169	12.1	0.73	April	C
-3.91	12.4	91.9	19.9	< 0.85	3.26	1.89	321	98.8	0.51	July	
-3.97	12.6	107	25.8	< 0.85	2.80	4.04	249	93.8	0.11	none	D
-3.84	33.5	111	19.3	0.97	3.40	5.53	167	91.0	0.14	none	E

Figure 7.17: In order to display the complex spatiotemporal situation, measuring results from April to July 2010 are combined with interpretations of the vertical solute distribution patterns and redox state classifications. A redox-anomaly was observed in well C, when the water table reached its lowest level end of April. Figure caption continued on next page.

Figure 7.17, continued: Later on, the water table increased by ~1.55 m until July (monsoon season), but remained unconfined. To this time, water from layer II became visible in well B and As concentrations close to the maximum appeared in well C. Data from wells A, D and E are average values (April-July 2010), while results for wells B and C are the exact values from 26/04/10 and 30/07/10. All concentrations in mg L^{-1}, except for $\delta^{18}O$ values (‰ VSMOW), the As(III) percentage (%), and As and U concentrations ($\mu g\ L^{-1}$).

7.3   DISCUSSION

7.3.1 SEDIMENT STRATIGRAPHY

Sequence stratigraphic interpretations based on the characterisation of OM, as well as on geochemical and lithological results, lead to similar conclusions as for the low As site (see APPENDIX II, Table A 2.6) and results of SARKA et al. (2009). An important difference is that the lithology includes here an additional facies F1, a clayey layer met in 39.2 m bls (see Figure 7.1). Increased TOC contents in facies F2 (between 32.0 and 39.2 m bls) provide additional information regarding the sedimentation environment of this depth range. Results allude to the conclusion that OM in sediments from F2 includes influences of terrestric matter that dates back to the early and middle Pleistocene. During this time, fluvial deposit filled the incised valleys that were eroded during the lowstand of the LGM at about 20 ka BP (see chapter 3.2). Embedded intrusions of mixed and poorly ordered muddy and gravelly sediments indicate high transport energies allowing a rapid reallocation of older sediments. Gravel grains are poor to medium rounded and consist of secondary carbonate nodules, which have likely formed during previous weathering and soil genesis. Increased TOC and mixed signals of freshwater POM and C_3 plants indicate that sediments of F2 originate from a freshwater swamp dominated environment.

Increased OM contents met in F2 and F4 result either from a high primary production and a following rapid burial, or from pronounced preservation effects within the sediments that prevented biodegradation (HOEFS 2009). Strongly biodegraded OM, which was reported from the Chakdah site, implies the former (ROWLAND et al. 2006). Another important question in this context is whether sedimentary OM is autochthonous (for example from marsh or swampland deposit) or allochthonous in nature. Poorly ordered sediments in F2 reflect pronounced relocation processes that suggest an allochthonous origin of the OM, whereas sediments in F3 and F4 are considered as undisturbed sequences with autochthonous OM.

7.3.2 *ROLE OF THE SURFACE AQUITARD*

Similar to the low site, the unconsolidated aquifer sediments are covered by 3.35 m thick clayey and silty layers of facies F4, which are interpreted as fluvial overbank deposits of the nearby Hooghly River. As discussed previously (chapter 6.3.2), these layers form an important surface aquitard, which prevents infiltration of oxygen- and nitrate-rich surface water and allows the establishment of moderate to strongly reducing conditions in the shallow aquifer below. Each time the water table reaches the clayey and silty aquitard in about 3.35 m bls (during the monsoon season), conditions change from unconfined to confined. Visible aggregations and mottles of secondary Mn- and Fe-(oxyhydr)oxides between 2.05 and 4.65 m bls are suitable markers for the position of the fluctuating unsaturated and oxic zone, while the permanently reducing zone is indicated by grey coloured sands that appear from 5.95 m bls on.

The pronounced As enrichment in F4 of up to 122 mg As kg^{-1} indicates a high retention potential of the sediments in this part, which results from the presence of Mn- and Fe-(oxyhydr)oxides as well as high contents of smectite and kaolinite, which all can adsorb significant amounts of As(III) and As(V) (MOHAN & PITMAN 2007). Hence, As and other trace elements like Zn, Ni, Cu are supposed to be removed by clay minerals and secondary formed Mn- and Fe-(oxyhydr)oxides as soon as anoxic groundwater rises into the aerobic vadose zone via capillary rise. In fact, sedimentary As is here predominantly associated with Fe-(oxyhydr)oxides according to SEP results. In previous studies, only little attention was paid to the surface near unsaturated zone. HARVEY et al. (2006) argue that these widespread, iron-rich sediments are a potential source of As, but present results reveal that this part is highly dynamic and acts as more like an effective sink for As rather than a source.

In addition, clinker fragments reaching up to a depth of two metres reflect a strong anthropogenic influence close to the surface. Hence, this part was artificially raised and/or reworked in the recent past, while sediments below are considered as undisturbed.

7.3.3 *ARSENIC IN SEDIMENTS*

Since geochemistry and mineralogy of the sediments are highly similar to the low As site, it is not surprising that Fe-(oxyhydr)oxides were identified as dominating host for As, too. In contrast to the As-enriched clayey and silty layers of facies F4, As contents in the sandy aquifer sediments (F3b, F3a and F2) are typically low. In contrast to the low As site, As contents are also accompanied by relatively increased TOC contents. The SEP further indicates that significant amounts of the sedimentary As are associated with OM, although this last extraction step also includes As-sulphides and recalcitrant As-minerals. This was previously suggested by AKAI et al. (2004) for sediments from Bangladesh, where As was found to be associated with OM besides Mn- and/or Fe-(oxyhydr)oxides and sulphides. A more precise characterisation was not possible since strong analytical limitations arose from the throughout low As contents.

7.3.4 GROUNDWATER PROPERTIES

Hydrological context. In order to explain the pronounced As enrichment in local groundwater, it is mandatory to identify past and present processes on the hydrochemical composition of the groundwater, which are highly similar to the low As site (see 6.3.4). Results allude to the conclusion that the fate of As is here also closely linked to the biogeochemical cycling of OM and Fe-(oxyhydr)oxides. The predominance of As(III) further points at an influence of DARPs. Vertical distribution patterns of key parameters (major solutes and redox sensitive compounds) reflect that the groundwater body is highly differentiated and that the situation is much more complex than at the low as site.

Hydrochemical stratification. The vertical distribution pattern of major solutes including conservative tracers Cl^-, K, $\delta^{18}O$ and $\delta^{2}H$ reflects the presence of two distinctive hydrological layers within the monitored part of the groundwater body. The boundary is assumed to be located around 25-26 m bls, in between the screening positions of wells B and C (see chapter 7.2.2.3). Stable isotopic values constantly plot below the LMWL, which is characteristic for evaporative influences (GAT 1996). Increased salinity as well as higher $\delta^{18}O$ and $\delta^{2}H$ values indicate that water of layer I was more affected by evaporation than of layer II. Mol ratios for conservative Cl^- and K in wells B and C were used to calculate enrichment factors for the two layers. Calculated ratios $[Cl^-_{well\ B} / Cl^-_{well\ C}]$ and $[K_{well\ B} / K_{well\ C}]$ are both 3.56, which further supports the presumed evaporative enrichment. Enrichment ratios of Ca, Mg and Na are pronounced lower in groundwater from layer I (1.19 – 1.51), expressing the non-conservative character of these elements.

Redox zoning. Vertically deviant distributions of NH_4^+, SO_4^{2-}, Fe and Mn further display a decreasing redox potential with depth and the manifestation of deviant redox zones (BORCH et al. 2010). Such a redox zoning is considered as characteristic for arsenic-affected shallow aquifers in the BDP and other parts of Asia (VAN GEEN et al. 2008). Redox sensitive trace elements like U, Mo and As form characteristic distribution patterns, too. These distinctive redox zones spatially superimpose both layers, which are supposed to have formed in the following. According to the detailed spatial characterisations done in Chakdah, these zones can be best described as plumes with vertical extensions of several hundred metres and thicknesses of a few decimetres (MÉTRAL et al. 2008). The development of such plumes requires intensive microbial redox activity and therefore degradable organic matter as well as an overlying clay aquitard layer, which prevents inflow of fresh surface water rich in O_2 and NO_3^-.

According to the redox classification of JURGENS et al. (2009), groundwater in wells A and B is primarily in state of Fe(III)/SO_4^{2-} reduction. A further distinction between Fe(III) and SO_4^{2-} reduction was not possible, because no H_2S determination was done. In the three wells below, SO_4^{2-} is depleted and methanogenesis possibly the prevailing redox process. Despite this, the availability of weakly ordered Fe-minerals could basically support Fe(III) reduction. In addition, fresh groundwater smelled of H_2S during the pumping experiment, clearly indicating SO_4^{2-} reduction.

Anomaly in well C. In well C, total Fe and Fe(II) concentrations are distinctively lower than in the other wells above and beneath. Since SO_4^{2-} and Ba concentrations are lower, too, precipitation of Fe- and Ba-sulphides must be considered in this specific depth range. This anomaly is persistent according to results of the entire monitoring. Considering characteristic SO_4^{2-} concentrations of layer I, the amount of possibly precipitated sulphides is in range of several milligrams per kilogram, which is too low neither to be visible in distribution of sedimentary TS contents, nor by respective XRD spectra.

In well C, highest As concentrations occurred, which are clearly de-coupled from dissolved Fe. As opposed to the well-known assumption that Fe-sulphides incorporate As (KIRK et al. 2010), O'DAY et al. (2004) reported that As(III) may not be incorporated in sulphate-low environments like redox transition zones when precipitation rates are fast. This results in an enrichment of As, if adsorption is inhibited, what could be related to high PO_4^{3-} concentrations.

Additionally, formation of new As binding sites by precipitation and transformation of residual Fe-phases is suppressed when dissolved Fe is removed during Fe-sulphide formation (see chapter 6.3.6).

Effects of lithology. Once different redox zones have established below the clayey and silty aquitard, the hydrochemically altered groundwater slowly drifts through the aquifer. Several lateral transitions in the grain size distribution are expected to cause a vertical anisotropy in groundwater flow, which is characteristic for the Bengal Basin and allows the preservation of hydrochemical deviant layers within the groundwater body (MICHEAL & VOSS 2009a,b). By comparing the position of the five monitoring well screens to the lithology and estimated K values, groundwater is considered to be drawn from three different parts of the aquifer during sampling (Figure 7.18). Wells A and B are expected to deliver from a medium sand rich strata located between 17 to 25 m bls, which is confined by silty fine sand. The well screen of well C is located in a medium sandy fine sand part, while wells D and E primarily deliver from between 32 and 38 m bls. The delivery rate of the submersible sampling pump (2.4 L min^{-1}) is expected to be low enough to prevent a disturbance of the deviant layers during sampling. An interpretation of the data is provided in Figure 7.19.

Adjacent pond. The isotopic composition of sampled pond water reflects a strong influence of evaporation, which is dedicated to the on-going dry season. Since the last recharge during the monsoon season, two months had passed (December 2009). Relatively high Fe, As and As(III) concentrations and an absence of NO_3^- surprisingly indicate anoxic and presumably reducing conditions in the pond water. Reducing conditions arising from a lack of O_2 are attributed to the small pond size (approximately 7 x 4 m with a depth of 2 m) and the water temperature of 24°C. A refill of the pond with groundwater by local inhabitants could be ruled out. Results of NEUMANN et al. (2009) suggest that pore waters in pond sediments are highly reducing and that As is released either by reductive dissolution of Fe-(oxyhydr)oxides and/or by DARPs.

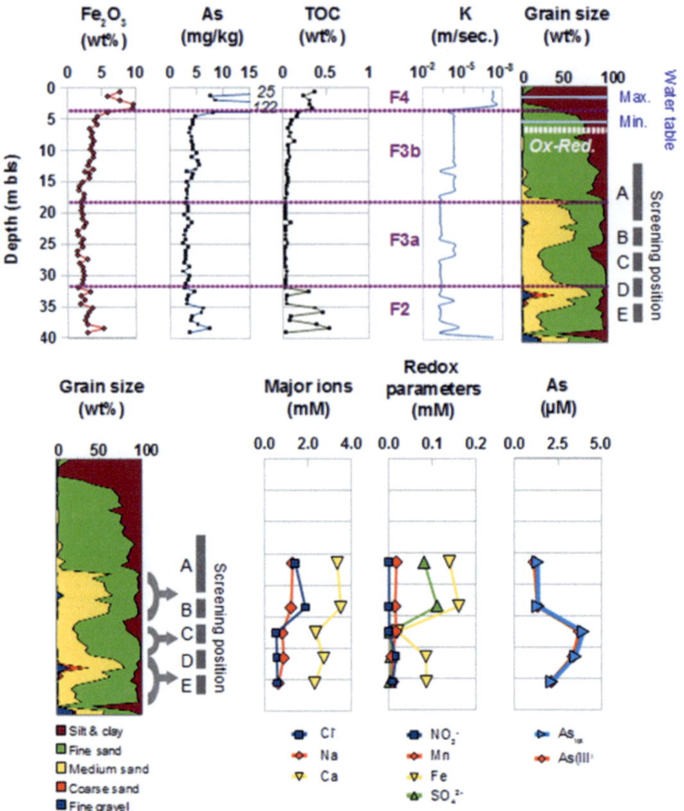

Figure 7.17: Comparison of sedimentary Fe, As and TOC contents, facies boundaries, estimated hydraulic conductivities (BEYER 1964), grain size distribution, position of the well screens and hydrochemistry of the five monitoring wells (samples from 03/12/09). In the grain size distribution included are the lowest (April) and highest water level (December) and the visible redox boundary in the sediment colour.

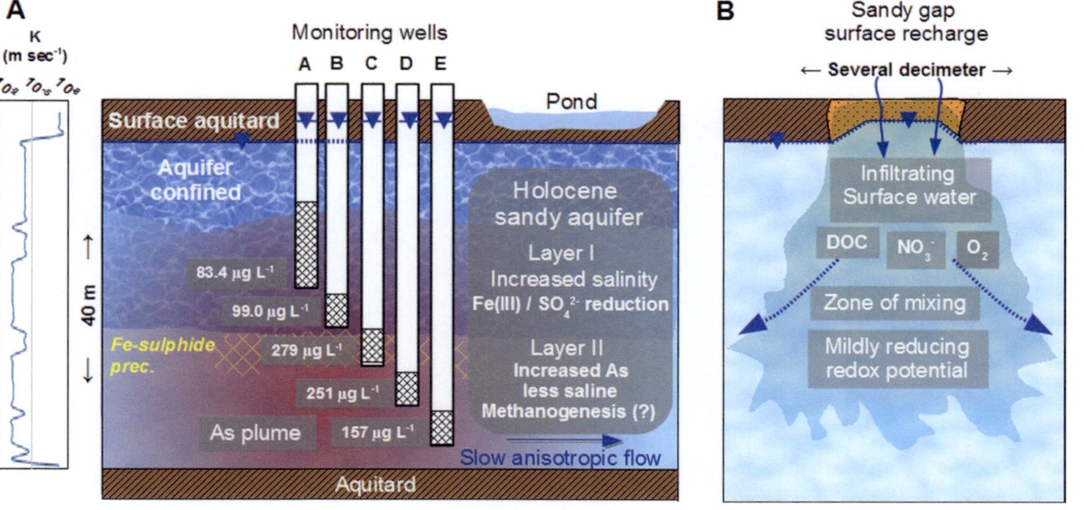

Figure 7.19: A) Schematic sketch of the aquifer architecture in the investigation area. Given As concentrations originate from samples taken on 03/12/09, when the aquifer was confined. It remains unclear if layer I really reached up to the aquitard since the well screen begins in 12 m bls. Additionally, the vertical extent of the As plume (indicated by red colour) and the zone of Fe-sulphide precipitation remain unknown. B) Illustration how infiltrating fresh surface water prevents a decline of the local redox potential below sandy lenses adjacent to the study site.

7.3.5 INFLUENCES OF HYDROSTATIC HEAD OSCILLATIONS ON THE LOCAL DISTRIBUTION OF ARSENIC

7.3.5.1 The groundwater extraction experiment

Change of experimental focus. The abstraction experiment was originally performed to induce temporary oscillations of the water table in order to aerate reducing parts of the aquifer. It was the intention to check whether induced changes in the local redox conditions influence the mobility of As. Since the experiment was conducted in the post-monsoon season, the aquifer was recharged and confined. During pumping, hydrostatic heads in the monitoring wells subsequently decreased until a steady state had adjusted. However, the applied delivery rate of 9 m^3 h^{-1} was not sufficient to create a virtue decrease of the water table and the intended temporary aeration of the saturated aquifer sediments was therefore not achievable.

The experimental focus was therefore modified and the interpretation aims at assessing short- and medium-term influences on dissolved As concentrations by locally restricted groundwater abstraction. In the Nadia district, private irrigation wells are widespread and typically confined to a depth of 18 m (NATH et al. 2008), which is highly similar to the pumping well A (well screen in 12-21 m bls). Additionally, the submersible pump that was used during the experiment was a common type that is sold and used for irrigation in West Bengal. Thus, the experiment is in the following interpreted as a simulation of irrigation pumping during the dry post- and pre-monsoon seasons.

Estimation of affected area. Groundwater was redirected from its preferred lateral flow paths towards pumping well A. Mixing of vertically and laterally drawn groundwater caused remarkable changes in the hydro-chemistry of respective monitoring wells. The total volume of the extracted water was about 1,700 m^3. Assuming a homogenous aquifer with a constant thickness of 36 m and an average pore volume of 30 %, each m^2 at the surface is equivalent to an approximate storage volume of 10.8 m^3 groundwater in the aquifer below. Hence, the extracted groundwater volume would be equivalent to an area of about 160 m^2. However, the actual size of the delivery area can be considered as substantially larger, since the confined water storage around the study site was permanently recharged by inflowing water. In addition, inhomogeneities in the virtue lithology cause spatially deviating storage capacities and hydraulic conductivities further enlarge the sphere of influence.

Mixing effects in wells A and B. Pronounced changes manifested in the hydrochemical compositions of groundwater from pumping well A and the adjacent observation well B, which were previously situated in layer I and in a less reducing redox zone (see chapter 7.2.2.2). This was primarily related to physical mixing as indicated by pronounced decreases in conservative tracers Cl$^-$, K, δ^{18}O and δ^2H. Most other major groundwater constituents (e.g., Mg, Ca and SO$_4^{2-}$) acted very similar and are therefore considered as conservative-like in context of the experiment. This behaviour is attributed to the relatively short rest cycles of 24 h between two pumping cycles, but also from the homogenous geochemistry and mineralogy within the respective aquifer sediments.

The comparison of samples taken immediately after 48 h of continuously pumping (13/12/09) with samples taken after the rest cycle (14/12/09) reflect slight influences of slowly and horizontally inflowing groundwater from layer I as indicated by changes in δ^{18}O, δ^2H and K values. Since the initial baseline values of K were characteristic for both layers, concentration changes were used to estimate the growing influence of groundwater from layer II in range of wells A and B (Figure 7.11). At the end of the experiment, water delivered from well B originated to more than >90 % from layer II, and about 75 % in pumping well A. Hence, pumping had pronounced effects on the vertically layered hydrochemistry of the upper two wells.

Influences on wells C, D and E. At a first glance, effects on groundwater around the well screen of well C appeared to be surprisingly weak. In respective samples, only minor changes occurred in trace element concentrations with deviant distributions (like Fe, Mn, and U). Only Li and PO_4^{3-} showed pronounced changes related to the excessive groundwater withdrawal. The composition of the sample taken directly after 48 h of pumping further revealed changes in several additional parameters ($\delta^{18}O$, δ^2H, Ca and Mg). This is attributed to a rapid lateral inflow of uninfluenced groundwater in this specific part of the aquifer once pumping was stopped, although the lithology does not indicate an increased hydraulic permeability.

The well screen is located close to those of wells B and D with distances of each 1.00 m in vertical and about 3.50 m in lateral direction. Hence, it remains unclear why the slightly increasing Fe concentrations remained at a comparatively low level, and increased Mn and Mo concentrations declined only marginal during the entire experiment.

Within the deepest wells D and E, most major and trace elements remained unaffected or showed minor changes (e.g., Cl^-, $\delta^{18}O$ and δ^2H). The underlying clay layer is expected to have prevented upstream of water from deeper parts and allowed only lateral inflow and mixing with groundwater of a similar hydrochemical composition.

Deviant behaviour of PO_4^{3-}, MO and DOC. Phosphate concentrations increased as only parameter exponentially in all five monitored wells immediately after the first pumping interval. This was similar to the sucrose injection at the low As site, where PO_4^{3-} increased as well after circular pumping. In addition, the sample taken directly after 48 h (13/12/09) of continuous pumping from well A revealed that PO_4^{3-} and Mo were much higher than in the sample taken after the rest cycle (14/12/09).

This behaviour clearly exceeded pure physical mixing, which is suitable to explain changes in conservative and most other solutes. Surface adsorption and desorption are the only potential processes that can explain these short-termed trends as described in the following.

As previously discussed for the low As site (chapter 6.3.5), increased PO_4^{3-} concentrations in groundwater can originate from microbial decomposition of OM as well as from reductive dissolution of hosting Fe-(oxyhydr)oxides, which was previously reported from oxbow sediments (LEWANDOWSKI & NÜTZMANN 2010). According to thermodynamic calculations, $H_2PO_4^-$ was the predominating species. In addition to strong adsorption via ligand exchange, negatively charged $H_2PO_4^-$ can further bind electrostically to positively charged surfaces. During the pumping experiment, loosely bound autochthonous $H_2PO_4^-$ became entrained by the strong artificial water flow created by pumping. Such short-termed mobilisation of PO_4^{3-} was reported from streams during flood events, when the flow velocity rapidly increased (DORIOZ et al. 1989, REDDY et al. 1999). During the rest cycle, increased $H_2PO_4^-$ concentrations decreased in pumping well A, which is attributed to a fast re-adsorption. The same applies to Mo (predominating: MoO_4^{2-}), which also declined rapidly in the sample from well A after the rest cycle of 24 h.

At the end of the experiment, DOC concentrations had manifold in all five wells, too (Table 7.6). Similar to PO_4^{3-}, this increase is attributed to desorption of poorly adsorbed autochthonous and electrically charged organic molecules. Despite this, a potential import of DOC with vertically and horizontally attracted groundwater during pumping as it was argued in other studies (e.g., HARVEY et al. 2002, NEUMANN et al. 2009) is considered unlikely in this situation.

Behaviour of arsenic. Evolving As concentrations in groundwater of the upper wells A and B correlated negatively with conservative tracers like Cl⁻ or $\delta^{18}O$. This conservative-like behaviour of As alludes to the conclusion that net increases originated from mixing with arsenic-enriched groundwater from layer II and not from an additional mobilisation. According to the comparison of samples taken before and after the rest cycle, neither short-termed As adsorption nor desorption appeared within all five wells (see Figure 7.10).

In wells D and E, As concentrations remained throughout stable. Despite this, As concentrations with As(III) as prevailing species had increased in groundwater from wells A, B and C to concentrations that were higher than the initial maximum of 296 $\mu g\ L^{-1}$ (well C, 03/12/09). According to the vertical distribution pattern of As that superimposes layers I and II, the initial As maximum of approximately 335 $\mu g\ L^{-1}$ was concealed between the well screens of wells C and D and was therefore not visible. This assumption is further supported by the following regular monitoring, which describes the mid-term effects of the experiment (chapter 7.3.5.2).

Influences of the adjacent pond. There is an intense debate whether ponds contribute to As release in the BDP or not (e.g., SENGUPTA et al. 2008 in contrast to NEUMANN et al. 2009). First, a potential influence requires a hydraulic connection between the pond and the aquifer below. During drilling of the monitoring wells in May 2008, the pond was completely dried out and it was noticed that the pond is sealed by clayey sediments. During the pumping experiment, there was no influence of pond water on the groundwater composition in the surface near filtered wells visible. Despite the extensive groundwater subtraction from the central well A, the water table fell by approximately 5 cm in 16 days. This is dedicated to evaporation during the throughout warm and dry weather, whereas a hydraulically connection between the pond and the aquifer below is therefore considered as unlikely.

7.3.5.2 Temporary changes during the monitoring

The question arises, which mechanisms that have formed the distinctive distribution pattern of As are still active and what were the consequences of the pumping experiment. The pumping experiment is considered as the key for interpretation of temporary changes that occurred during the monitoring (see chapter 7.2.2.3). It is shown that trends in the groundwater composition of wells A, B and C were closely attributed to vertical and horizontal movement of the groundwater body. A brief summary of the processes following the pumping experiment is presented in form of a conceptual model in Figure 7.20.

Return to hydrochemical baseline following the pumping experiment. During the dry pre-monsoon phase between January and April 2010, groundwater compositions in the monitoring wells A and B subsequently returned to conditions similar to the initial situation before the abstraction experiment. Hence, the well screens were relocated into layer I, which caused a decline in As concentrations and an increase in conservative Cl^- and $\delta^{18}O$. This is attributed to a slow lateral inflow of unperturbed groundwater. From January on, an additional vertical downward movement of the hydrochemically stratified water body appeared relative to the static well screen positions. As a result, the groundwater hydrochemistry within the monitoring wells gradually returned towards the initial composition that was observed before the pumping experiment (Figure 7.20). This process lasted, regarding As, between two (central pumping well A) and six weeks (well B). The temporal difference is attributed to differences in the lateral flow velocities and the heightened position of the well screen of well A.

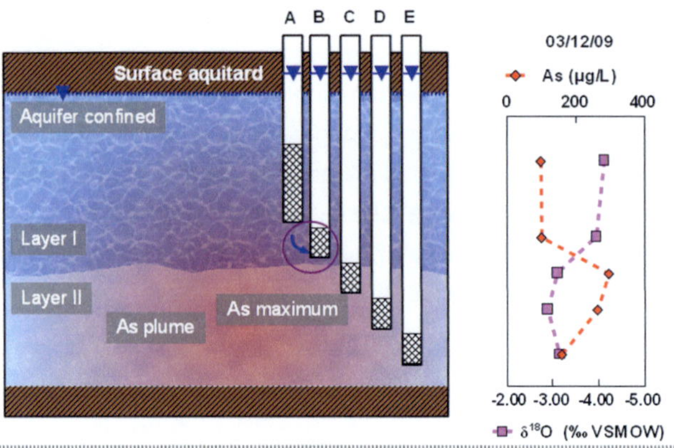

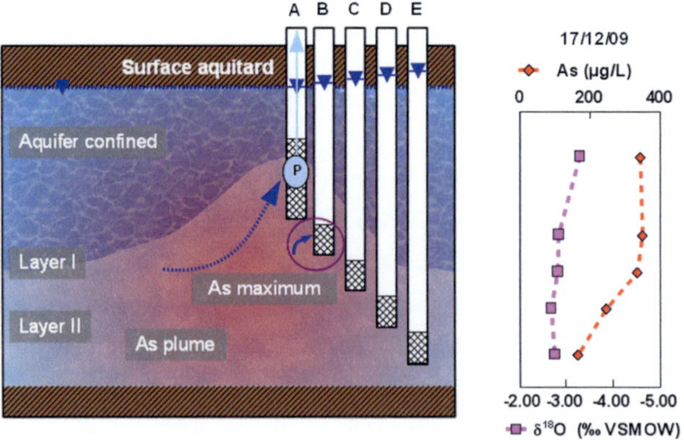

Figure 7.20, part I: Schematic sketch illustrating the assumed position of the two layers (indicated in the depth profile by $\delta^{18}O$) and the As plume before (A) and in different seasons after the pumping experiment (B, C, D). During the following dry pre-monsoon season, the water table subsequently decreased, while the hydrochemistry returned to its initial situation (C). Increased evaporation induced capillary rise, which alluded to precipitation of secondary Fe-(oxyhydr)oxides within the surface aquitard. Figure continued on next page.

C) Pre-monsoon dry season, groundwater table decreasing in unconfined aquifer
→ wells A and B again low As + increased salinity

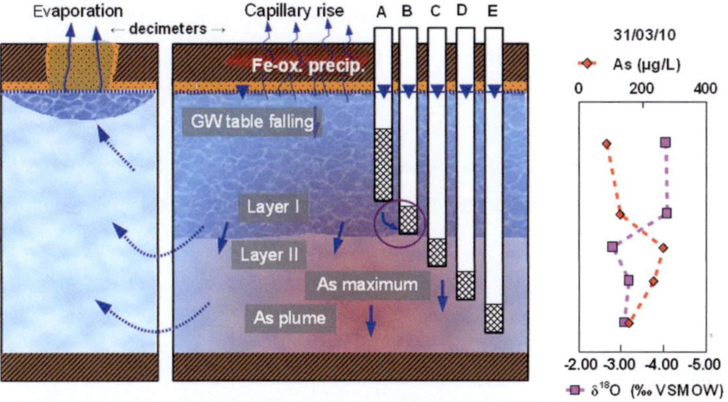

D) Monsoon season, gradual vertical shift of water body + As plume and less saline layer
→ well B now high As + low salinity in contrast to well A

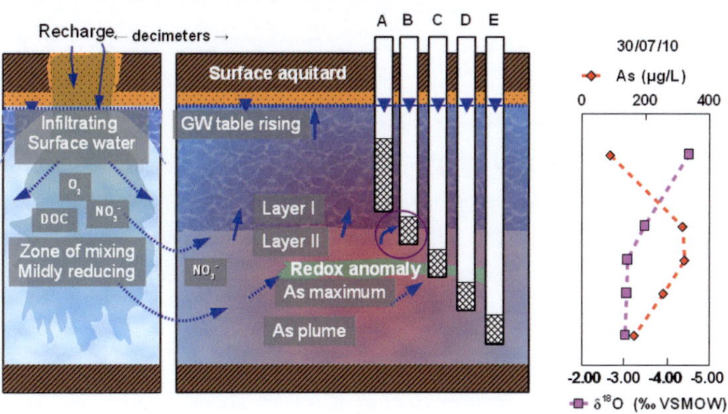

Figure 7.20, part II: During the monsoon rains, the aquifer was recharged via surface water inflow at adjacent sandy lenses in the surface aquitard, which resulted in displacement of groundwater at the high As site (D). The water table rose, and the As plume as well as layer I consequently moved upwards and the layer boundary successively passed the well screen of well B.

Hydrochemical compositions in groundwater of wells D and E primarily remained constant, only the strongly increased PO_4^{3-} concentrations constituted an exception. Phosphate concentrations rapidly declined in all wells and stabilised at concentrations, which were slightly higher than the baseline. This is another evidence for the fast re-adsorption ability of PO_4^{3-}, which did not cause any visible As release. At the end of the dry pre-monsoon season, the groundwater table had reached the lowest level and the baseline values were restored in nearly all parameters and wells.

Parallels to 2009. Although the installation of the monitoring wells was completed in June 2008, the central well A collapsed soon and was rebuilt until December 2008. Close before the first sampling, the well was developed according to common practice, comprising intensive flushing of the well screen. This is done by strong pumping for several hours, which is highly similar to the abstraction experiment. Even the point of time and therefore the hydrological conditions were nearly identical. Hence, it is not surprising that changes in hydrochemical parameters during the following weeks were highly similar (see figure 7.12).

Trends in As during both pre-monsoon seasons are attributed to mixing of groundwater with deviant As concentrations, which originated from pumping. Interestingly, the initial As maximum that occurred directly at the beginning of the monitoring in well B (387 µg L^{-1}) was not reached during the abstraction experiment in 2009 (349 µg L^{-1}).

Temporal trends arising during the monsoon season 2010. Trends in As and other characteristic parameters occurred in well B during the monsoon season in 2010, but not in 2009 (see Figures 7.12 and 7.14). Again, changes were obviously controlled by movement of the stratified groundwater body, which shifted about 1.55 m upwards during the monsoon season (Figure 7.20 D). As a result, reverse temporal trends manifested in As and other characteristic parameters in monitoring samples of well B, displaying the complete vertical layer boundary between end of April (layer I visible) and July (layer II visible).

Similar vertical movement of the water column appeared during the monsoon season in 2009, too, but hydrochemical parameters including As remained at constant levels. This decisive difference between 2009 and 2010 is affiliated to consequences of the pumping experiment. The boundary between the saline and arsenic-low layer I and the less saline, but arsenic-rich layer II appeared to be coincidentally located around the well screen position of well B. The abstraction experiment than has caused a slight upward shift of the vertical layer boundary within the water column, which was obviously sufficient to relocate the screening of well B into the layer below during the following monsoon season. This local disturbance in the water body was most likely retained by the slow and lateral groundwater flow, which prevented a fast removal of the disturbed groundwater.

The character of groundwater started to change at the end of April (hydrostatic head at 5.23 m bls, aquifer unconfined) before it reached a stable condition in July (hydrostatic head at 3.68 m bls, aquifer still unconfined). Hence, the approximate thickness of this layer boundary was to this time 1.55 m, assuming a stable boundary that moved parallel to the hydrostatic head.

Arsenic behaviour in well C. Pronounced changes in the hydrochemistry of well C during 2010 can be similarly explained. During the monsoon season 2010, concentrations of As, As(III) and Fe strongly increased after they had reached a minimum at the end of the dry pre-monsoon season (Figures 7.12 and 7.15). These temporary trends are considered as an expression of the distinctive vertical distribution patterns of these solutes, which generally superimpose layers II and I. As discussed previously, highest As concentrations occur in the depth range around the well screen of well C. In contrast, other parameters like $\delta^{18}O$, Cl⁻ or K were characterised by invariant concentrations within layer II and in range of wells C, D and E, and behaved therefore indifferent towards the vertical upward shift of the water table. When the lowest level of the water table was reached at the end of the pre-monsoon season (April 2010), As(III) was not the predominating As species although As concentrations remained generally increased. In addition, nearly no Fe occurred in solution to this time while dissolved U concentrations showed a maximum. This is surprising, since U mobility is sensitive to redox state changes and solubility normally decreases by reduction to U(IV) in reducing environments (BORCH et al. 2010). Additionally, the presumed zone of iron-sulphide precipitation is also located in this range. All this points at the presence of a thin layer where the redox state of Fe, As and U is obviously influenced by unknown processes. The reason for this "redox anomaly" remains obscure, but throughout stable $\delta^{18}O$ values exclude a potential influence of oxygen-rich recharge water.

Arsenic increase in well A. Since the beginning of the regular monitoring, the comparatively low initial As level in groundwater from well A rose slightly, but continuously. The same applies to PO_4^{3-}, whereas the As/PO_4^{3-} mol ratio remained nearly constant. A release of additional As via PO_4^{3-}-As-exchange is therefore excluded. According to the strong correlation between As, Fe and conservative $\delta^{18}O$, this increase can be rather attributed to an enduring shift of the hydrochemical layers inside the water body than to active As release. The reason for this constant shift remains unclear, but clear pumping of the well after drilling in December 2008 disturbed the local hydrological context. The same applies to the time after the pumping experiment in 2009, when concentrations of Cl^-, SO_4^{2-}, K, and Fe showed first decreasing, than increasing trends parallel to the oscillating water table. These fluctuations are again considered to result from deviant vertical distribution patterns of these elements, which hold maximum concentrations around the well screen of well B.

A contrasting effect was reported by MC ARTHUR et al. (2010) from a monitoring well at the 40 km east located JAM study site. Here, As concentrations declined together with Fe, Cl^-, Ca and SO_4^{2-} (PO_4^{3-} was not included). This was attributed to flushing with younger recharge water.

The arsenic – iron – phosphate - system. Throughout high positive correlations between As, PO_4^{3-} and Fe appeared in the monitoring samples. As previously discussed for results of the field survey and the low As site, positive correlations between As and PO_4^{3-} support the assumption that both are controlled by the same release mechanism, which was previously proposed by MC ARTHUR et al. (2001) and VAN GEEN et al. (2004). This further supports the hypothesis postulated in chapter 6.3.4 (f), whereas As and PO_4^{3-} were previously released during reductive dissolution of Fe-(oxyhydr)oxides and that all available binding sites are currently occupied in the meanwhile matured aquifer system. Specific mol ratios occurred in the monitoring wells between these three solutes, which only changed in well B during the layer shifts as previously discussed. The ratio of Fe to PO_4^{3-} in well C additionally reflects a loss of Fe, since the regression line has a comparable slope than the other wells, but does not cross the zero point (see Figure 7.13). This loss of dissolved Fe supports the assumed precipitation of Fe-sulphides as discussed in chapter 7.3.4.

Further hydrochemical processes affecting arsenic mobility. The pumping experiment has demonstrated that the prevailing hydrochemical system with its vertical stratified water column has a strong tendency to return to the previous, undisturbed condition. Temporal changes in the hydrochemical composition of the monitoring wells are either related to seasonal changes in the groundwater table, or to effects of pumping (2008: clear pumping of well A; 2009: pumping experiment).

The hydrochemical situation is partly highly similar to the low As site, where As release is an active process. In contrast to the low As site, additional release of As into local groundwater did here not occur. As discussed in chapter 6.3.4, microbial processes are considered as key players in the genesis of arsenic-enriched groundwater in the investigation area. The situation at this site reflects that the aquifer is rather matured and near the state of equilibrium than a young and highly dynamic system. Once released, abiotic reactions (namely competitive adsorption) and prevailing hydrology (inducing vertical and lateral transport within the aquifer) determine the further fate of As.

8. SYNTHESIS

- CONCEPT OF ARSENIC MOBILISATION AND DISTRIBUTION IN THE STUDY AREA

8.1   COMPARISON OF THE TWO STUDY SITES

Geomorphology and lithostratigraphy. Both study sites are situated in the floodplain area of the BDP in the Nadia District, only 3.1 km apart from each other. The sites are located in a distance of about 8.2 km (high As site) and 11.0 km (low As site) east of Chakdah city, which is situated at an elevated interfluve of the Hooghly River. The Hooghly River flows from north to south and represents the only active watercourse within the study area. Numerous abandoned river beds indicate frequent channel changes in the recent past (chapter 5, Figure 5.1). In accordance to the surface geomorphology, sediments at both sites draw a picture of a highly dynamic sedimentation history that tells about basin-wide and complex tectonic processes as well as past climatic changes (see chapter 3.2 and sequence stratigraphic interpretation summarised in Figure 8.1). Aquifer sediments are primarily composed of fine and medium sands with varying proportions of silt. Principal minerals are quartz, feldspars, carbonates and mica. In addition, clay minerals (smectite and potentially traces of kaolinite and illite) occur in the clayey and silty surface aquitard. Although As contents in the sandy aquifer are generally low at both sites, they are still capable to distinctively increase the groundwater concentration, even if released only in parts (Figure 8.1). This assumption is supported by the fact that Fe-(oxyhydr)oxides (amorphous, weakly ordered as well as crystalline ones) represent the prevailing host for As in the sediments. In the surface aquitard, where secondary Fe-(oxyhydr)oxides are supposed to periodically precipitate together with associated trace elements during the dry season, a partly pronounced enrichment of As appears.

Groundwater evolution. Groundwater compositions reflect that microbially mediated As mobilisation by Fe(III) and As(V) reduction occurred at both sites, which is indicated by simultaneous enrichments of As(III), PO_4^{3-} and Fe(II) (chapters 6.2.2.1 and 7.2.2.1). Active Fe(III) reduction and concomitant As(III) release was observable in well A of the low As site during the monitoring and in the course of the biostimulation experiment (chapters 6.2.2.2 and 6.2.2.3). Additionally, short- and mid-term concentration changes in dissolved As and PO_4^{3-} following the biostimulation experiment support the assumption that As release was accompanied by competitive adsorption to residual Fe-(oxyhydr)oxides with empty binding sites (fast) and newly formed Fe-minerals that formed by transformation and precipitation (slow). Generally stable vertical hydro-chemical layers observed during the first year of monitoring indicate that the local groundwater flow is slow and anisotropic as assumed by MICHAEL & VOSS (2009a,b). Subsequent changes in the local groundwater compositions that followed both in-situ experiments lasted several weeks, further supporting this assumption. Slow groundwater flow enables subsequent accumulation of As when release is an active process as described for well A at the low As site.

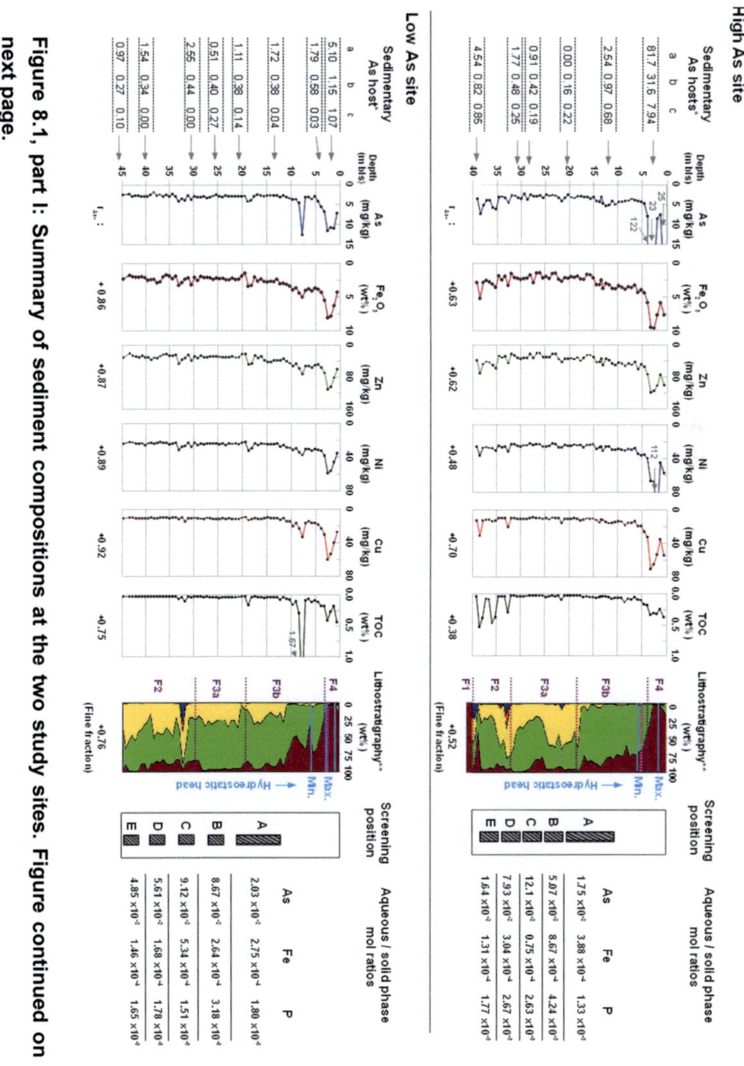

Figure 8.1, part I: Summary of sediment compositions at the two study sites. Figure continued on next page.

* **Extractable As (mg/kg) in sequential extraction fractions:**

a: Surface adsorbed As
(fractions I and II)

b: Arsenic co-precipitated with amorphous + weakly ordered Fe-(oxyhydr)oxides
(fractions III and IV)

c: Arsenic co-precipitated with crystalline Fe-(oxyhydr)oxides
(fraction V)

** **Sequence stratigraphy:**

F4:
Delta-floodplain progradation sequence;
Hooghly overbank deposits, clay and silt.
Age: Middle to late Holocene

F3b:
Transgressive system tract sequence;
Estuarine channel fill sediments, fine to middle sand with increasing silt share.
Age: Holocene, < 9 ka BP

F3a:
Transgressive system tract sequence;
Fluvial channel fill sediments; fine and middle sand.
Age: Holocene, < 9 ka BP

F2:
Lowstand system tract sequence;
fluvial incised valley fillings following LGM;
Fine and medium sand with gravel lenses and silt.
Age: early Holocene

F1:
Lowstand system tract sequence;
Sequence boundary marking LGM (?);
Marine clay or clayey overbank deposits.
Age: early Holocene / late Pleistocene

Figure 8.1, part II: Sequential extraction fractions and sequence stratigraphy.

Differences between the two study sites. Despite the comparable geochemical compositions of the aquifer sediments (Figure 8.1), distinctive differences were noticeable regarding the spatiotemporal distribution of As in groundwater of the two sites:

• The high As site comprises a more complex hydrochemically stratified water column and pronounced higher dissolved As concentrations than the low As site. Since mineralogy and lithology of the aquifer sediments are highly similar, differences in the hydrochemistry are attributed to processes of groundwater evolution. However, differing As concentrations originate obviously not from Mn(IV) reduction, since Mn concentrations are similar in all 10 monitoring wells.

Increased alkalinities and partly increased Fe concentrations in groundwater point at a more intensive microbial activity at the high As site, which is considered as major reason for the stronger accumulation of dissolved As, which is at both sites predominantly As(III).

Furthermore, a redox anomaly was observed around the depth range of well C at the high site, where unexpected high As concentrations of up to 296 µg L^{-1} occurred.

• Although As concentrations have constantly increased in groundwater at both study sites during the monitoring, the underlying mechanisms differ. In case of the low As site, increasing As concentrations in the shallowest well A are attributed to active mobilisation by microbial Fe(III) reduction (see chapter 6.3.6). Here, on-going As release was additionally superimposed by seasonal changes in concentrations of As in the groundwater, which induced pronounced increases during the dry seasons and decreases at the beginning of the monsoon rains.

At the high As site, a constant increase in the As concentration occurred in the central well A, which was accompanied by changing δ^{18}O values, while dissolved Fe remained nearly constant (see chapter 7.3.5.2). The groundwater abstraction experiment further caused fluctuations in well B (high As site), which had obviously enduring influences on the position of the hydrochemical boundaries. Both temporal effects were obviously linked to an increasing hydrostatic head (net increase up to 2.10 m) and therefore to recharge of the aquifer during the monsoon rains. Hence, the conclusion is drawn that both increases in As concentrations were caused by shifts in the hydrochemically stratified water column and not by active As release.

8.2 IMPLICATION FOR ARSENIC MOBILISATION AND ACCUMULATION

In the following, a concept is developed that describes the spatio-temporal distribution of dissolved As in groundwater of the BDP.

Groundwater evolution in the Bengal Delta Plain. The hydro-chemistry in the investigation area generally reflects a pronounced influence of alteration of the groundwater chemistry through numerous biotic and abiotic processes and subsequent reactions. Although shallow groundwater in the BDP is in depths down to 100 m relatively young (<100 a, AGGARWAL et al. 2000, STUTE et al. 2007), hydrochemical parameters reflect intensive influences of recent groundwater evolution that is closely linked to microbially mediated decomposition of buried OM under anaerobic conditions (ROWLAND et al. 2006 & 2007). As a result of microbial Fe(III) reduction, the water-sediment system is not in equilibrium and the water is supersaturated in respect of Fe-(oxyhydr)oxides (e.g., hematite, magnetite, siderite and vivianite, depending on present E_H, pH and hydrochemical composition). Compared to other reactions that influence the hydrochemical composition of groundwater, microbially mediated redox processes are comparatively fast, but require a continuous supply with OM. If no additional input of OM and TEA like O_2 and NO_3^- occurs, the redox potential in the local aquifers will gradually decrease with time (and depth), until the pool of degradable organic carbon is exhausted. Different redox zones with distinctive hydrochemical layers form with time, where the presence of a widespread surface aquitard prevents inflow of fresh and potentially oxygen- and nitrate-rich surface water (MÉTRAL et al. 2008).

Arsenic release and enrichment. Microbially mediated redox reactions are the key mechanisms for As mobilisation, especially through Fe(III) reduction. This was demonstrated by the biostimulation experiment (see chapter 6.2.2.2) as well as by monitoring results of well A at the low As site (chapter 6.2.2.3). The predominance of As(III) in all monitoring wells additionally raises the question if DARPs were involved in As mobilisation. Since many FeRB carry the *arrA*-gene (OREMLAND & STOLZ 2005, ZOBRIST et al. 2000), it is not possible to distinguish between As release via reductive dissolution of hosting Fe-(oxyhydr)oxides and As(V) reduction (KAPPLER 2011). It is important to note that no high initial sedimentary As contents are required to create problematic As concentrations in groundwater (see calculation in chapter 2.2.2).

The characteristically slow and preferred vertical groundwater flow entails in the following an enrichment of As(III) and other compounds that are released into groundwater like HCO_3^-, PO_4^{3-}, Fe(II) and Mn(II), while a flush-out of arsenic-enriched groundwater into the Bay of Bengal is extremely delayed. Arsenic release via oxidative weathering of arsenic-bearing sulphides (HARVEY et al. 2002) during aeration of primarily submerged reduced sediments was not observed during the monitoring and field experiments. Sedimentary As is primarily associated with Fe-(oxyhydr)oxides and monitoring results showed no significant impact of groundwater oscillations on As release.

Shallow wells of the investigation area. In order to compare outcomes of the monitoring to those of the field survey, average values obtained during the monsoon season 2009 are used to exclude the previously described seasonal changes. During this time, As concentrations in shallowest monitoring wells (wells A and B high As site, well A low As site) were elevated, but still pronounced lower than in the deeper monitoring wells (Table 8.1).

The same is true for shallow tube wells of the investigation area with depths below 20 m. In about 20 m depth, the redox boundary between NO_3^- /Mn(IV) and Fe(III)/As(V) reduction manifests in a sharp increase of As and Fe concentrations in groundwater. Shallow wells still contain SO_4^{2-} and are considered to be primarily dominated by Fe(III) reduction and concomitant As release as observed in well A of the low As site. Some samples from shallow tube wells with depths <20 m are enriched in principal solutes, especially Cl⁻, while Na/Cl ratios are lower compared to deeper wells. The increased salinity is expected to originate from mixing with evaporation influenced water. This assumption is supported by $\delta^{18}O$ and $\delta^{2}H$ values of wells A and B at the high As site.

Obviously, Mn(IV) reduction has occurred and created problematic Mn concentrations in groundwater, which are in excess of the Indian drinking water threshold value of 0.30 mg L^{-1} (if no other drinking water source is available, otherwise: 0.10 mg L^{-1}; IS 10500) in all monitoring wells and in 45.4 % of the local tube wells. Only five shallow survey samples were found to be in state of Mn(IV) reduction, which means Fe(III) reduction was not reached (see Figure 5.3 in chapter 5.2). Manganese-oxides are important potential hosts for As, too (BORCH et al. 2010), but their abundance in the sediments is pronounced lower compared to Fe-(oxyhydr)oxides. Indeed, As concentrations in samples that are in state of Mn(IV) reduction are clearly lower (<10 µg L^{-1}) than in samples reflecting Fe(III)/SO_4^{2-} reduction or methanogenesis. This strengthens the assumption that Mn(IV) reduction plays a minor role in As release, which was previously concluded from the biostimulation experiment and from the hydrochemistry of the study sites.

Table 8.1, part I: Average concentrations and standard deviations in groundwater of the two study sites (between May to November 2009). Table continued on next page.

Well (m bls)	TA (mg/L)	Cl^- (mg/L)	SO_4^{2-} (mg/L)	PO_4^{3-} (mg/L)	Fe (mg/L)	Mn (mg/L)	As* (µg/L)	As/PO_4^{3-} (mol ratio)	Na/Cl (mol ratio)	$\delta^{18}O$ (‰)**
High As site (n: 12)										
A (12-21)	545 (± 25)	58.7 (± 6.5)	10.5 (± 1.44)	2.20 (± 0.24)	3.43 (± 0.46)	0.75 (± 0.06)	75.3 (± 5.7)	0.04 (± 0.00)	0.63 (± 0.07)	-3.21 (± 0.36)
B (22-25)	551 (± 37)	58.6 (± 7.2)	10.3 (± 2.71)	2.43 (± 0.23)	7.31 (± 0.74)	0.52 (± 0.10)	90.4 (± 9.4)	0.05 (± 0.01)	0.63 (± 0.05)	-3.33 (± 0.50)
C (26-29)	490 (± 24)	22.9 (± 3.3)	<0.85	2.69 (± 0.28)	1.38 (± 0.26)	0.80 (± 0.07)	281 (± 14)	0.14 (± 0.01)	1.35 (± 0.15)	-3.90 (± 0.15)
D (30-33)	514 (± 10)	19.1 (± 1.1)	<0.85	2.67 (± 0.28)	4.01 (± 0.36)	0.29 (± 0.02)	214 (± 8)	0.11 (± 0.01)	1.50 (± 0.13)	-4.15 (± 0.30)
E (34-37)	473 (± 18)	20.6 (± 1.6)	<0.85	2.70 (± 0.20)	4.39 (± 0.32)	0.57 (± 0.02)	134 (± 5)	0.07 (± 0.01)	1.07 (± 0.11)	-3.86 (± 0.17)
Low As site (n: 11)										
A (12-21)	445 (± 24)	7.59 (± 0.78)	<0.85	1.68 (± 0.25)	3.05 (± 0.33)	0.60 (± 0.08)	36.1 (± 3.8)	0.03 (± 0.00)	3.53 (± 0.34)	-4.42 (± 0.17)
B (24-27)	443 (± 5)	2.43 (± 0.28)	<0.85	2.65 (± 0.24)	3.18 (± 0.14)	0.43 (± 0.03)	127 (± 8)	0.06 (± 0.00)	7.95 (± 1.02)	-4.43 (± 0.23)
C (30-33)	445 (± 17)	2.42 (± 0.26)	<0.85	2.21 (± 0.23)	3.31 (± 0.45)	0.42 (± 0.26)	112 (± 10)	0.07 (± 0.00)	7.85 (± 0.90)	-4.38 (± 0.12)
D (36-39)	436 (± 10)	2.47 (± 0.26)	<0.85	1.97 (± 0.09)	2.78 (± 0.16)	0.42 (± 0.04)	108 (± 4)	0.07 (± 0.00)	7.53 (± 0.93)	-4.41 (± 0.13)
E (42-45)	441 (± 1)	2.36 (± 0.25)	<0.85	1.82 (± 0.10)	2.26 (± 0.14)	0.60 (± 0.04)	106 (± 6)	0.08 (± 0.00)	7.76 (± 0.77)	-4.56 (± 0.11)

*As(III) percentage for the monitoring wells between 91.3 and 97.2 %
**VSMOW

Table 8.1, part II: Results of the field survey. Results are divided into three classes based on the tube well depth. To provide an overview of the large data set of the field survey, median values and quartiles (25 % and 75 %, in brackets) are given.

Well	TA	Cl⁻	SO_4^{2-}	PO_4^{3-}	Fe	Mn	As*	As/PO_4^{3-}	Na/Cl	$\delta^{18}O$
(m bls)	(mg/L)	(mg/L)	(mg/L)	(mg/L)	(mg/L)	(mg/L)	(µg/L)	(mol ratio)		(‰)**
					Field survey (n: 165)					
<20 m (n: 37)	nd	19.3 (10.7 to 33.7)	6.83 (<0.85 to 13.4)	0.43 (0.11 to 1.74)	1.71 (0.18 to 3.39)	0.33 (0.21 to 0.48)	17.0 (3.1 to 38.4)	0.03 (0.01 to 0.06)	1.35 (1.01 to 2.28)	nd
20–40 m (n: 90)	nd	15.8 (7.8 to 32.3)	5.09 (<0.85 to 6.14)	3.06 (1.21 to 4.23)	3.81 (2.25 to 6.24)	0.31 (0.20 to 0.39)	70.6 (42.6 to 88.6)	0.02 (0.01 to 0.04)	1.66 (1.12 to 2.91)	nd
>40 m (n: 36)	nd	3.78 (2.52 to 5.39)	<0.85 (<0.85)	0.14 (0.06 to 0.46)	0.46 (0.05 to 1.52)	0.14 (0.08 to 0.28)	14.9 (1.50 to 82.2)	0.15 (0.02 to 0.26)	7.36 (4.77 to 11.4)	nd

*As(III) percentage not determined for tube well samples
**VSMOW

The zone of highest arsenic concentrations. According to the field survey, As accumulation is highest in depths of about 20 to 40 m, where moderate to strong reducing conditions prevail (see chapter 5.3.1). The same distribution pattern was reported from the neighbouring country Bangladesh, where the British Geological Survey (BGS) and the Department of Public Health Engineering (DPHE) analysed 3,534 wells local wells during a comprehensive field survey (published in 2001, see Figure 8.2). Monitoring wells of the high and low As study sites located in this depth range (wells B to E) showed throughout constant As concentrations. Hence, the conclusion may be drawn that As release into groundwater is completed in this part of the aquifer, is in equilibrium with re-adsorption, or is extremely slowed down so that increases are not visible within the one year of monitoring.

Wherever As concentrations in groundwater are increased, PO_4^{3-} concentrations are high (>1.00 mg L^{-1}), too. They are both primarily released during microbial Fe(III) reduction and accumulate where the groundwater flow is slow and anisotropic. The behaviour of dissolved As and PO_4^{3-} following the two in-situ experiments demonstrated that an increase in the PO_4^{3-} concentrations could not induce an additional release of As into groundwater via PO_4^{3-}-As-exchange as proposed by ACHARYYA et al. (1999).

Affected depth ranges and intensities of As release generally depend on the quantity as well as the quality of available OM, since As mobilisation is primarily influenced by metabolic activities of anaerobic microbes capable of Mn(IV), Fe(III), As(V), and SO_4^{2-} reduction. Based on the biostimulation experiment, the hypothesis was postulated that an increase in the PO_4^{3-} concentrations increases in turn the enrichment of As in groundwater by reducing the availability of adsorption sites. This mechanism was entitled as "competitive adsorption" (see chapter 6.3.4, f).

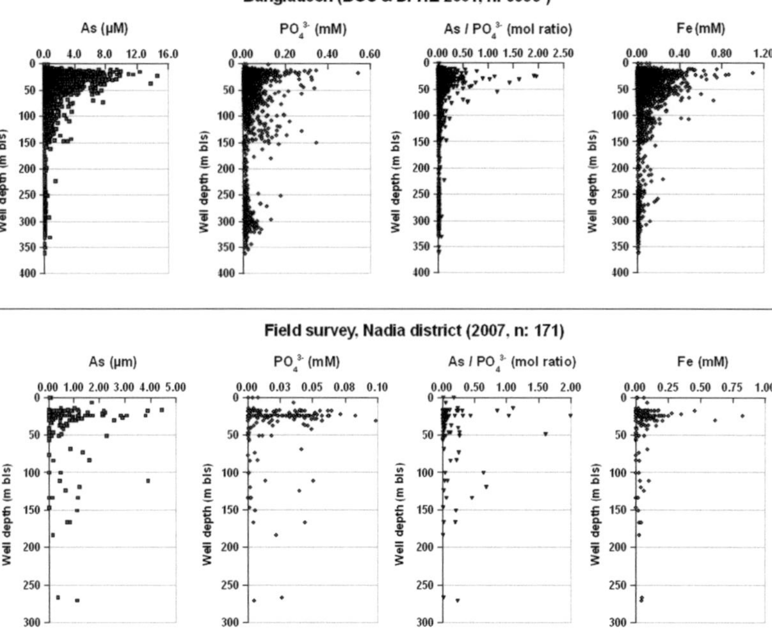

Figure 8.2: Depth plots of As, PO$_4^{3-}$ and Fe concentrations in samples from the field survey in comparison to results of the BGS and DPHE* survey in Bangladesh. In Bangladesh, concentrations peak in similar depths as in the field survey, supporting the assumed position of the redox boundary in about 20 m and the zone of highest As enrichment in about 20 to 40 m depth. In Bangladesh, no explanation for the PO$_4^{3-}$ increases in wells deeper than 250 m was found, but due to the high groundwater ages, fertilisers were excluded as potential source (BGS & DPHE 2001).

*One outlier removed (24 µM As)

Deep wells. In wells deeper than ~40 m, concentrations of Fe, Mn, As, and PO_4^{3-} are lower than in the active redox zone located above. This means arsenic was not released or concentrations of arsenic slowly declined after release. A decline can be the result of a flush-out into the Bay of Bengal after As release stopped. An interruption of As release occurs as soon as the pool of mobilizable sedimentary As is exhausted, or available organic matter that fuels microbially mediated As release is depleted.

The mechanism of competitive adsorption offers a new alternative explanation. Groundwater dating and hydrochemical flow models have demonstrated that groundwater age in the BDP generally increases with depth (MICHAEL & VOSS 2009a, MUKHERJEE et al. 2011, STUTE et al. 2007). In addition, Na/Cl ratios in the survey samples demonstrate that processes of groundwater evolution continue with time. This ratio can be considered as a suitable indicator to assess the intensity of water-rock interactions that shaped present groundwater and is therefore a useful tool to estimate the degree of groundwater maturity in the BDP. Furthermore, long residence times facilitate the kinetically slow precipitation of super-saturated mineral phases (e.g., siderite, vivianite) and/or transformation reactions of present Fe(III) oxides through dissolved Fe(II) (e.g., formation of magnetite). Hence, As and PO_4^{3-} can be retained by newly formed binding sites (see chapter 6.3.4 for underlying hypothesis) and their concentrations will slowly decrease when the maturity of the groundwater increases over time and/or with depth. Indeed, absolute concentrations of As and PO_4^{3-} are highest in a range of about 20 to 40 m bls and decrease in deeper wells (Table 8.1 and Figure 8.2).

The same applies to the BGS & DPHE data set. Here, a flush-out of dissolved solutes in these deeper aquifer parts can also not be excluded, but groundwater ages of up to 3 ka in 200 to 300 m depth support the hypothesis of mineral precipitation (AGGARWAL et al. 2000).

The higher the concentrations of competing PO_4^{3-}, the more As should remain in groundwater since PO_4^{3-} exhibits higher binding affinities under neutral pH values (DIXIT & HERING 2003). Other than expected, As/PO_4^{3-} mol ratios do not clearly increase with depth (Figure 8.2). This deviant behaviour is attributed to the existence of other binding mechanisms for PO_4^{3-} that do not compete with As(III) and/or As(V) adsorption. For example, the pumping experiment revealed that PO_4^{3-} anions can also be weakly bound by electrostatic adsorption.

The presumed removal of dissolved Fe with time further explains why some of the field survey samples (12 out of 16) were found to be in a mixed redox state (chapter 5, Figure 5.3). In these samples, Fe concentrations subsequently fell below the threshold value (0.10 mg L^{-1}, JURGENS et al. 2009). By contrast, field survey samples without SO_4^{2-}, but significant amounts of NO_3^- further indicate mixing with percolating groundwater in a less reducing redox state. This is attributed to extensive groundwater abstraction (see chapter 5.3.1). In addition, the abstraction experiment performed at the high As site demonstrated that mixing can also cause strong increases of As concentrations in low-arsenic groundwater.

Thus, groundwater of the investigation area was and is controlled by various, interconnected biotic and abiotic processes (see Table 8.2) that create and preserve the characteristic bell-shaped depth profile of dissolved As (HARVEY et al. 2002). However, anthropogenic pumping can change this specific distribution pattern. The hydrochemical architecture of aquifers with arsenic-enriched groundwater in the floodplain of the BDP is illustrated in Figures 8.3 and 8.4.

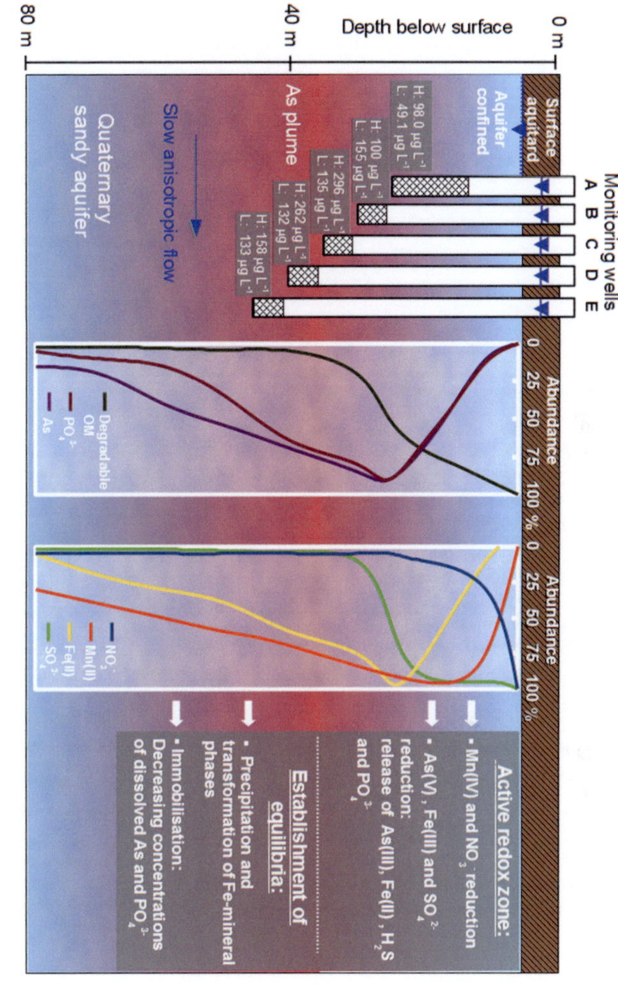

Figure 8.3: Schematic sketch of the distribution of redox sensitive elements and compounds in arsenic-enriched aquifer parts of the investigation area. Highest As concentrations appear in about 20 to 40 depth. H: high As site, L: low As site; included As concentrations from 03/12/09.

213

Table 8.2: Summary of partly counteracting processes that have influenced the groundwater chemistry in the investigation area.

Process:	Detectable in:	Impact on groundwater chemistry:
Mixing with evaporation influenced water	Wells A and B at the high As site	Increase in heavier ^{18}O and ^2H isotopes, increased salinity.
Carbonate weathering	All groundwater data, biostimulation experiment	Increased alkalinity and Ca, Mg, Ba and Sr concentrations; Circum-neutral pH.
Ion exchange	All groundwater data	Na-Ca-exchange causes the increase of Na/Cl ratios with depth and time.
Microbial redox reactions: NO_3^- reduction Mn(IV) reduction Fe(III) reduction As(V) reduction SO_4^{2-} reduction	All groundwater data, biostimulation experiment	Increased alkalinity, intensification of carbonate weathering; Establishment of specific redox zones with respective decrease in OM, NO_3^- and SO_4^{2-} and increases of Mn(II), Fe(II), As(III) and PO_4^{3-} concentrations.
Precipitation of supersaturated Fe-mineral phases and transformation of weakly ordered Fe-phases by Fe(II)	All groundwater data, biostimulation experiment	Decreasing Fe(II) concentrations with depth / time.
Competitive (re-)adsorption	All groundwater data, biostimulation experiment	Depletion of available binding sites; Simultaneous increase of the dissolved As and PO_4^{3-} concentrations, followed by a decrease of As and PO_4^{3-} concentrations with depth and time.
Anthropogenic influences: Mixing caused by pumping and/or input of OM	Field survey data, abstraction experiment, biostimulation experiment	Pumping causes disturbances of the naturally occurring stable hydrochemical layers and As distributions; Inflow of OM into reducing zones may trigger fast microbially mediated As release.

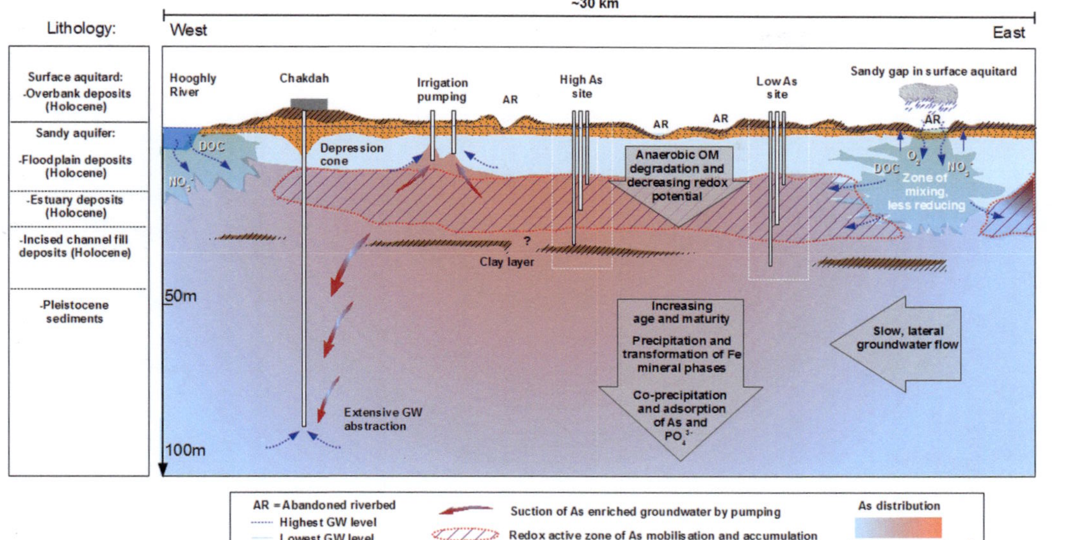

Figure 8.4: Overview of processes involved in mobilisation, enrichment and distribution of As concentrations in groundwater of the investigation area. Recharge of the aquifer occurs, wherever sandy lenses in the aquitard allow infiltration of fresh water (according to MÉTRAL et al. 2008). The figure also indicates that the high As site is situated in the centre of an As plume (after CHARLET et al. 2007). The redox anomaly in range of well C at the high As site was not considered. Groundwater abstraction generally influences both, shallow (see results of the pumping experiment, chapter 7.3.5) as well as deeper parts of the groundwater (NATH et al. 2008).

Conditions causing arsenic enrichment in groundwater.

The following specific local conditions need to be fulfilled that As can be released / accumulated in shallow groundwater:

- Permanently low redox conditions in the range of Fe(III) and As(V) reduction, which requires in turn:

  - Availability of degradable OM that controls the intensity and duration of microbial metabolic reactions;

  - Limited amounts of dissolved O_2 and NO_3^- that allow decline of the redox potential;

  - A protective surface aquitard that prevents inflow of TEA and oxygen-rich water;

  - Presence of arsenic-bearing Fe-(oxyhydr)oxides in the sediments;

  - Increased water temperatures to accelerate active microbial metabolic reactions;

- Slow horizontal groundwater flow that entails subsequent enrichment of dissolved As;

- High concentrations of PO_4^{3-} to inhibit or slow down the adsorption of dissolved As.

Due to the highly variable sediment morphology of the floodplain system, most of these parameters are not homogenously distributed in groundwater and the aquifer sediments. Hence, the intensity of microbial activity and the concomitant As release differs from site to site, resulting in a heterogeneous distribution (both horizontally and vertically) of the As concentrations in the groundwater. This became evident at the two study sites, where the groundwater chemistry was shaped by similar processes, but the concentrations of As in groundwater strongly differed in certain depth intervals.

Temporal changes in arsenic concentrations. In addition to the spatial heterogeneity of As enrichments in groundwater, monitoring results from two wells (well B at the high As site and well A from the low As site) further revealed pronounced temporary changes in As concentrations. Temporary changing dissolved As concentrations were previously reported from the BDP (MC ARTHUR et al. 2010), but never with such a high temporal resolution. These changes arose from mixing and/or vertical movement of the groundwater body and embedded hydrochemical layers that are controlled by seasonal oscillations of the hydrostatic head. In well A of the low As site, these seasonal changes superimpose a constant increase in As concentrations arising from active Fe(III) reduction. Since the average depth of private tube wells is around 24 m in the area (see chapter 5.2), many wells are considered to tap water from vertically layered redox zones, where As concentrations can seasonally vary.

Anthropogenic effects. Distribution patterns of dissolved As in local tube wells indicate intensive groundwater abstraction via deep governmental wells causing drawdown of younger, less mature and often arsenic-enriched groundwater. This explains why increasing As concentrations were observed in previously low-arsenic deep wells (e.g., FENDORF et al. 2010, WINKEL et al. 2011) and also why 34 field survey samples were classified as a mixture of deviant redox states, holding moderate amounts of NO_3^- as well as high concentrations of Fe and As (chapter 5.3.1).

The pumping experiment proved that this mechanism works in the opposite direction, too. Here, simulation of extensive abstraction of shallow groundwater demonstrated a high vulnerability of groundwater towards mixing with arsenic-enriched groundwater.

Irrigation with shallow groundwater in times of the dry season is common practice in the fertile floodplain east of Chakdah. In 1998/1999, about 60 % of the 8,450 ha fields have been irrigated with shallow ground-water (NATH et al. 2008).

Together with numerous private and governmental tube wells for drinking water gain, a significant disturbance of the local complex hydrochemistry can be expected by re-directing the natural groundwater flow in the respective area. Induced effects on the composition of water delivered by local tube wells can be short-termed, but the abstraction experiment demonstrated that also long-lasting shifts of the layer boundaries can occur, alluding to an increase in As concentrations in the respective wells.

Although water abrasion by itself does not cause release of additional amounts of arsenic, suction of DOC (e.g., derived from infiltrating sewage) into reducing aquifer zones may potentially induce a rapid stimulation of indigenous microbes and a concomitant mobilisation of As. Microorganisms can overcome periods of carbon absence by reducing the metabolic activity or by formation of endospores, until fresh OM (re-)activates their metabolism (KIEFT et al. 2007). This was previously discussed by several authors (e.g., HARVEY et al. 2002, NEUMANN et al. 2009 and POLYA & CHARLET 2009) and was also proved by the sucrose field experiment.

Hence, it could be shown that pumping may trigger unpredictable effects on the local groundwater properties and must be therefore recognized as an additional important factor that may influence the distribution of As in groundwater of the BDP.

9. CONCLUSIONS

Here, answers to the key questions raised in chapter 1 are provided:

1. How is As release linked to the availability of OM in the aquifer?

The As release itself is the result of naturally occurring and widespread microbially mediated decomposition of OM, which leads to the development of reducing conditions in the aquifer sediments. The hydrochemical monitoring documented (for the first time) an active mobilisation of As(III) by Fe(III) reduction in one of the monitoring wells. Results of the sucrose biostimulation experiment further support this assumption, but demonstrate that microbial mobilisation of As is accompanied by a concomitant im-mobilisation brought about the adsorption to residual and/or newly formed Fe-(oxyhydr)oxide phases.

Below the zone of As accumulation in about 20 to 40 m depth, groundwater is more evolved as compared to younger water from above. Due to exhaustion of OM, microbial processes slow down. As a result, the ground-water chemistry is strongly disturbed here and the water-sediment system is not in equilibrium as indicated by the prevailing SI at the two study sites. Long groundwater residence times generally foster the progress of kinetically slow processes like the precipitation of supersaturated Fe-minerals or Na-Ca-exchange, which increase with depth and time. As a result of Fe-mineral precipitation and transformation, new binding sites are formed, inducing the immobilisation of As- and PO_4^{3-}-oxyanions. Hence, concentrations of As slowly decline when the maturity of the groundwater further increases with time.

2. Which biogeochemical reactions are induced by indigenous microbial communities that control the mobilisation of As into local groundwater?

The in-situ biostimulation experiment demonstrated at the low As site the important role of geomicrobiological processes in water-sediment inter-actions. Results of this experiment were combined with outcomes of the hydrochemical monitoring and results of the high As site as well as the field survey to develop a concept that explains the spatial distribution pattern of dissolved As visible in groundwater of the investigation area. The As/Fe, As/PO$_4$$^{3-}$ and Fe/PO$_4$$^{3-}$ mol ratios have been proved to be helpful indicators to identify processes influencing the mobility of As in aquifers of the BDP. The highly similar behaviour of As- and PO$_4$$^{3-}$-oxyanions and the enrichment in groundwater indicates that both are mainly controlled by the same mechanism, which was identified as the reductive dissolution of Fe-(oxyhydr)oxides. Additionally, temporal changes induced by the biostimulation experiment point at the competitive adsorption of As and PO$_4$$^{3-}$ to available binding sites of old and newly formed Fe-minerals. This process occurs parallel to mobilisation and outlasts the release. Hence, a long-lasting net increase of As concentrations in groundwater occurs only, when the available binding sites are already occupied either by As or PO$_4$$^{3-}$. These results further confirm laboratory findings that demonstrated competing As release and retention via microbial degradation of OM.

3. What are the specific local conditions that determine As mobilisation, and how do such conditions arise?

The fate of As in aquifers of the BDP was and still is controlled by multiple and superimposed processes and factors (summarised in Figure 9.1), which have created distinctive spatiotemporal distribution patterns of dissolved inorganic As in the groundwater.

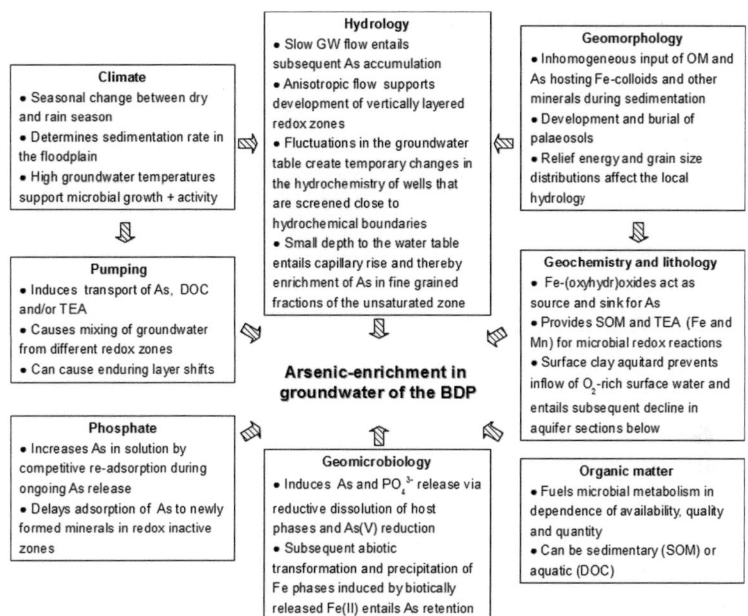

Figure 9.1: Complex interconnections between different factors that influence the As distribution and finally lead to As enrichment in shallow groundwater.

An important prerequisite for the mobilisation of As is a moderate to strongly reducing redox potential, which requires in turn degradable organic matter and a protective surface aquitard that prevents infiltration of oxygen- and nitrate-rich surface water. Another important aspect is the slow and preferentially horizontal groundwater flow, which enables subsequent enrichment of dissolved As after its mobilisation. High concentrations of dissolved PO_4^{3-}, which is also released during Fe(III) reduction, will further reduce As immobilisation by competitive adsorption. The availability of organic matter and the presence of sandy gaps in the surface aquitard are directly influenced by sedimentation processes and are inhomogeneously distributed within the BDP. As a result, vertical and horizontal distributions of As in groundwater may strongly vary within narrow margins.

4. How is the response of dissolved As to interfering abiotic processes like temporal changes in the groundwater level?

Generally, the subsequent accumulation of As in groundwater following microbial controlled mobilisation is, in turn, controlled by abiotic hydrological processes and hydrochemical equilibrium reactions (e.g., precipitation of oversaturated Fe-minerals and competitive adsorption of As and PO_4^{3-}). Since the hydrochemical equilibrium reactions are strongly influenced by prior microbial processes like the reductive dissolution of Fe-(oxyhydr)oxides, they are inextricably linked with each other. Hence, a clear separation of As release and accumulation into biotic and abiotic processes is not possible. It was further shown that pronounced temporal changes in As concentrations can occur, which were directly linked to seasonal fluctuations of the hydrostatic head.

Another important and new outcome is that the surface aquitard can act as an important sink for dissolved As, which is attributed to capillary rise and precipitation of Fe-(oxyhdr)oxides under oxic conditions.

Anthropogenic activities can locally interfere with and/or overprint the established distribution patterns of dissolved As. In the near future, the fast growing population will increase the demand for drinking water and irrigation. Due to the easy accessibility of shallow groundwater, groundwater abstraction will further increase the pressure on the groundwater resources of the BDP as well as of many other Asian floodplain and deltaic areas. Hence, mitigation strategies need urgently to be reappraised to provide drinking and irrigation water to the local population without endangering the strongly limited low arsenic water reserves.

OUTLOOK

In order to transfer the outcomes of this study to the whole BDP and other arsenic-affected Asian regions, it is necessary to investigate more than two single spots. This is important, since the controlling parameters of As mobility may often act in opposing directions. To get a more comprehensive picture, more investigations are necessary that focus on the new aspects described in this thesis. In order to save time and money, it would be convenient to apply the new concepts to already existing study sites like that of CHARLET and co-workers (Chakdah area), MC ARTHUR and his group (JAM area), or the areas in Bangladesh investigated by VAN GEEN and many others.

More attention needs to be paid on the following aspects related to groundwater As enrichment and mobility:

- **Determination of the influence of the surface aquitard.** The role of the overlying surface aquitards in the BDP as well as in other arsenic affected floodplain and deltaic areas of Asia remains unclear. More information is necessary on the distribution and spatial extent of the clayey and silty deposits and on the occurrence of reducing conditions in sub-surface aquitards (MÉTRAL et al. 2008). This could be a helpful tool to constrain and predict the occurrence of arsenic-enriched groundwaters. Another important aspect to be closer investigated is the role of these compact sediments as sink for dissolved As, when arsenic- and iron-rich groundwater enters the unsaturated zone via capillary rise and Fe-(oxyhydr)oxides precipitate.

Similar processes can be expected in the widespread rice paddies, where strongly reducing conditions may develop in the soil.

• **Monitoring of redox transition zones.** Monitoring results from the high As site demonstrate how dynamic and small scaled differences in local redox conditions can be. As a result, dissolved As, but also Mn, reach concentrations in reducing aquifers of the BDP that give reason to concern. Forthcoming studies, notably such related to risk assessment, should be aware and take into account that strong spatiotemporal variations in As concentrations may occur over time.

• **Characterisation of microbial communities and mineral interactions.** It is necessary to better investigate interactions between the input of reactive OM into reducing parts of the aquifer, the provoked activity of microbial communities, available Fe-minerals and the concomitant transformations of these minerals in assessing the mobilisation of As in natural aquifer systems. The identification of microbial strains in different redox zones and determination of active metabolic pathways by cutting-edge molecular-biologic methods in different redox zones is one of the most stringent and still pending issues, allowing to find out the common features of the arsenic-mobilising microbial communities.

A newly developed method is the use of biofilm in-situ microcosms, which can be directly introduced into groundwater (e.g., GILHAM et al. 1990). This provides the opportunity to incubate and trap microbes directly in their natural habitat. By varying the available growth medium, for example, different kinds of Fe-minerals, such microcosms could be used to trap and identify those FeRB that respire the respective mineral phases, and to observe the biogeochemical mediated transformation of Fe-minerals at the same time (CUMMINGS & MAGNUSON 2007).

• **Groundwater abstraction.** The groundwater abstraction experiment demonstrated that intensive abstraction from presumably arsenic-free groundwater bears the risk of attracting arsenic-enriched groundwater in direction of the pumping well. There is a pressing need to find alternative ways to provide safe drinking water before the last reserves of low-arsenic water are contaminated.

• **Arsenic - iron - phosphate interactions.** The hypothesis of As and PO_4^{3-} adsorption offers explanations why As remains in solution, and why dissolved As concentrations slowly decline with increasing depth. This concept requires further validation by laboratory column and batch experiments as well as by evaluation of previously published monitoring data from other affected areas (e.g., the Red River Delta in Vietnam).

Furthermore, if more detailed information was available regarding the sorption behaviour of As, interactions occurring during transport of arsenic- and phosphate-rich groundwater induced by anthropogenic pumping could be used to model reactive transport of As. This would allow to estimate the local vulnerability of low arsenic water reserves towards inflowing arsenic-increased groundwater.

REFERENCES

ABERNATHY, C.O., LIU, Y.-P., LONGFELLOW, D., APOSHIAN, H.V., BECK, B., FOWLER, B., GOYER, R., MENZER, R., ROSSMAN, T., THOMPSON, C., WAALKES, M. (1999): Arsenic: Health Effects, Mechanisms of Actions, and Research Issues. Environ. Health Perspect., 107, 593-597.

ACHARYYA, S.K., CHAKRABORTY, P., LAHIRI, S., RAYMAHASHAY, B.C., GUHA, S., BHOWMIK, A. (1999): Comment on Nickson et al. Arsenic poisoning of Bangladesh groundwater, Nature, 401, 545.

ACHARYYA, S.K., LAHIRI, S., RAYMAHASHAY, B.C., BHOWMIK, A. (2000): Arsenic toxicity of groundwater in parts of the Bengal basin in India and Bangladesh: the role of Quaternary stratigraphy and Holocene sea-level fluctuation. Environ. Geol., 39, 1127-1137.

ACHTNICH, C., BAK, F., CONRAD, R. (1995): Competition for electron donors among nitrate reducers, ferric iron reducers, sulfate reducers, and methanogens in anoxic paddy soil. Biol. Fertil. Soils, 19, 65-72.

ADRIANO, D.C. (2001): Trace Elements in Terrestrial Environments: Biogeochemistry, Bioavailability, and Risks of Metals, 2nd edition, Springer, New York.

AGGARWAL, P.K., BASU, A.R., KULKARNI, K.M., FROEHLICH, K., TARAFDAR, S.A., ALI, M., HUSSAIN, A. (2000): A report on isotope hydrology of groundwater in Bangladesh: Implications for characterization and mitigation of arsenic in groundwater. International Atomic Energy Agency, Project BGD/8/016. Available from: http://www-naweb.iaea.org/napc/ih/documents/other/BGD_Report.pdf.

AGGARWAL, P.K., BASU, A.R., KULKARNI, K.M. (2003): Comment on "Arsenic Mobility and Groundwater Extraction in Bangladesh" (I). Science, 25, 584.

AHMAD, S.A., BANDARANAYAKE, D., KHAN, A.W., HADI, S.A., UDDIN, G., ABDUL, M. (1997): Arsenic contamination in ground water and arsenicosis in Bangladesh. Int. J. Environ. Health Res., 7, 271-276.

REFERENCES

AHUJA, S. (2008): Solutions for Arsenic Contamination of Groundwater. In: AHUJA, S., Ed. (2008): Arsenic contamination of groundwater: Mechanism, Analysis, and Remediation. 367-376, John Wiley & Sons, Hoboken, New Jersey.

AKAI, J., IZUMI, K., FUKUHARA, H., MASUDA, H., NAKANO, S., YOSHIMURA, T., OHFUJI, H., ANAWAR, H.M., AKAI, K. (2004): Mineralogical and geomicrobiological investigations on groundwater arsenic enrichment in Bangladesh. Appl. Geochem., 19, 215–230.

ALAM, M.G.M., ALLINSON, G., STAGNITT, F., TANAKA, A., WESTBROOKE, M. (2002): Arsenic contamination in Bangladesh groundwater: a major environmental and social disaster. Int. J. Environ. Health Res., 12, 236-253.

ALAM, M.M., CURRAY, J.R., CHOWDHURY, M.L.R., GANI, M.R. (2003): An overview of the sedimentary geology of the Bengal basin in relation to their regional tectonic framework and basin-fill history. Sediment. Geol., 155, 179–208.

ALLISON, M.A., KHHAN, S.R., GOODBREAD JR., S.L., KUEHL, S.A. (2003): Stratigraphic evolution of the late Holocene Ganges-Brahmaputra lower delta plain. Sediment. Geol., 155, 317–342.

ANGEL, R.J. (1988): High-pressure structure of anorthite. Am. Min., 73, 1114-1119.

APPELO, C.A.J., POSTMA, D. (1996): Geochemistry, groundwater and pollution. A.A. Balkeme, Rotterdam, Brookfield.

APPELO, C.A.J., VAN DER WEIDEN, M.J.J., TOURNASSAT, C., CHARLET, L. (2002): Surface Complexation of Ferrous Iron and Carbonate on Ferrihydrite and the Mobilization of Arsenic. Environ. Sci. Technol., 36, 3096–3103.

ARGOS, M., KALRA, T., RATHOUZ, P.J., CHEN, Y., PIERCE, B., PARVEZ, F., ISLAM, T., AHMED, A., RAKIBUZ-ZAMAN, M., HASAN, R., SARWAR, G., SLAVKOVICH, V., VAN GEEN, A., GRAZIANO, J., AHSAN, H. (2010): A prospective cohort study of arsenic exposure from drinking water and all-cause and chronic disease mortality in Bangladesh. The Lancet, 376, 252-258.

BEDNAR, A.J., GARBARINA, J.R., BURKHARDT, M.R., RANVILLE, J.F., WILDEMAN, T.R. (2004): Field and laboratory Arsenic speciation methods and their application to natural-water analysis. Water Res., 38, 355-364.

BERG, M., TRAN, H.C., NGYEN, T.C., PHAM, H.V., SCHERTENLEIB, R., GIGER, W. (2001): Arsenic Contamination of Groundwater and Drinking Water in Vietnam: A Human Health Threat. Environ. Sci. Technol., 35, 2621–2626.

BERG, M., LUZI, S., TRANG, P.T.K., VIET, P.H., GIGER, W., STÜBEN, D. (2006): Arsenic Removal from Groundwater by Household Sand Filters: Comparative Field Study, Model Calculations, and Health Benefits. Environ. Sci. Technol., 40, 5567-5573.

BEYER, W. (1964): Zur Bestimmung der Wasserdurchlässigkeit von Kiesen und Sanden aus der Kornverteilungskurve. WWT, 14, 165-168.

BGS AND DPHE (2001): Arsenic contamination of groundwater in Bangladesh. Vol. 2: Final report. BGS Technical report WC/00/19, Vol. 2. Available from: http://www.bgs.ac.uk/arsenic/bphase2/reports.htm.

BHATTARACHARYA, P., WELCH, A.H., AHMED, K.M., JACKS, G., NAIDU, R. (2004): Arsenic in groundwater of sedimentary aquifers. Appl. Geochem., 19, 163-167.

BISH, D.L., VON DREELE, R.B. (1989): Rietveld refinement of non-hydrogen atomic positions in kaolinite. Clays Clay Min., 37, 289-296.

BISWAS, A., MAJUMDER, S., NEIDHARDT, H., HALDER, D., BHOWMICKA, S., MUKHERJEE-GOSWAMI, A., KUNDUA, A., SAHAA, D., BERNER, Z., CHATTERJEE, D. (2011): Groundwater chemistry and redox processes: Depth dependent arsenic release mechanism. Appl. Geochem., 26, 516-525.

BLAKE, R.L., HESSEVICK, R.E., ZOLTAI, T., FINGE, L.W. (1966): Refinement of the hematite structure. Am. Min., 51, 123-129.

BLAIR, N.E., CARTER JR., W.D. (1991): The carbon isotope biogeochemistry of acetate from a methanogenic marine sediment. Geochem. Cosmochim. Acta, 56, 1247-1258.

BORCH, T., KRETZSCHMAR, R., KAPPLER, A., CAPPELEN, P.V., GINDER-VOGEL, M., VOEGELIN, A., CAMPBELL, K. (2010): Biogeochemical redox processes and their impact on contaminant dynamics. Environ. Sci. Technol., 44, 15-23.

BUNDSCHUH, J., LITTER, M.I., NICOLLI, H.B., HOINKIS, J., BHATTARACHARYA, P. (2010): Identifying occurrences of groundwater arsenic in Latin America: A continent-wide problem and challenge. In: JEAN, J.S., BUNDSCHUH, J., BHATTARACHARYA, P., Eds. (2010): Arsenic in Geosphere and Human Diseases. Arsenic 2010: Proceedings of the 3rd Int. Cong. on Arsenic in the Environment. 512-516, CRC Press, Leiden.

REFERENCES

BURGES, W., HOQUE, M.A., MICHAEL, H.A., VOSS, C.I., BREIT, G.N., AHMED, K.M. (2010): Vulnerability of deep groundwater in the bengal aquifer system to contamination by arsenic. Nature Geosci., 3, 83-87.

CAMPBELL, K.M., MALASARN, D., SALTIKOV, C.W., NEWMAN, D.K., HERING, J.G. (2006): Simultaneous Microbial Reduction of Iron(III) and Arsenic(V) in Suspensions of Hydrous Ferric Oxide. Environ. Sci. Technol., 40, 5950–5955.

CHAPELLE, F.H. (1993): Ground-water microbiology and geochemistry. John Wiley & Sons Inc., New York.

CAPPUYNS, V., SWENNEN, R. (2008): The Use of Leaching Tests to Study the Potential Mobilization of Heavy Metals from Soils and Sediments: A Comparison. Water Air Soil Pollut. 191, 95-111.

CHARLET, L., ROMAN-ROSS, G., SPADINI, L., RUMBACH, G. (2003): Propagation of a natural arsenic plume in West Bengal, India. J. de Physique IV, 107, 285–288.

CHARLET, L., POLYA, D. (2006): Arsenic in Shallow, Reducing Groundwaters in Southern Asia: An Environmental Health Disaster. Elements, 2, 91-96.

CHARLET, L., CHAKRABORTY, S., APPELO, C.A.J., ROMAN-ROSS, G., NATH, B., ANSARI, A.A., LANSON, M., CHATTERJEE, D., BASU-MALLIK, S. (2007): Chemodynamics of an arsenic "hotspot" in a West Bengal aquifer: A field and reactive transport modelling study. Appl. Geochem., 22, 1273–1292.

CHATTERJEE, A., DAS, D., MANDAL, B.K., CHOWDHURY, T.R., SAMANTA, G., CHAKRABORTI, D. (1995): Arsenic in Ground Water in Six Districts of West Bengal, India: The Biggest Arsenic Calamity in the World Part I. Arsenic Species in Drinking Water and Urine of the Affected People. Analyst, 120, 643-650.

CHEN, C.J., CHEN, C.W., WU, M.-M., KUO, T.-L. (1992): Cancer potential in liver, lung, bladder and kidney due to ingested inorganic arsenic in drinking water. Br. J. Cancer, 66:888–892.

CHEN, S.-L., DZENG, S.R., YANG, M.-H., CHIU, K.-W., SHIEH, G.-M., WAI, M.W. (1994): Arsenic Species in Groundwaters of the Blackfoot Disease Area, Taiwan. Environ. Sci. Technol., 28, 877-881.

CHOWDHURY, T.R., BASU, G.K., MANDAL, B.K., BISWAS, B.K., CHOWDHURY, U.K., CHANDA, C.R., LODH, D., ROY, S.L., SAHA, K.C., ROY, S., KABIR, S., QUAMRUZZAMAN, Q., CHAKRABORTI, D. (1999): Arsenic poisoning in the Ganges delta. Nature, 401, 545-546.

CHRISTENSEN, T.H., BJERG, P.L., BANWART, S.A., JAKOBSEN, R., HERON, G., ALBRECHTSEN, H.-J. (2000): Characterization of redox conditions in groundwater contaminant plumes. J. Contam. Hydrol., 45, 165-241.

COKER, V.S., GAULT, A.G., PEARCE, C.I., VAN DER LAAN, G., TELLING, N.D., CHARNOCK, J.M., POLYA, D.A., LLOYD, J.R. (2006): XAS and XMCD Evidence for Species-Dependent Partitioning of Arsenic During Microbial Reduction of Ferrihydrite to Magnetite. Environ. Sci. Technol., 40, 7745-7750.

COPLEN, B.C., HERCZEG, A.L., BARNES, C. (2000): Isotope engineering – using stable isotopes of the water molecule to solve practical problems. In: COOK, P., HERCZEG, A.L., Eds. (2000): Environmental tracers in subsurface hydrology. 83, Kluwer Academic Publishers, Norwell.

CORNELL, R.M., SCHWERTMANN, U., Eds. (2003): The iron oxides. Structure, Properties, Reactions, Occurrences and Uses. 2nd edition. 438, Wiley-VCH, Weinheim.

COZZARELLI, I.M., WEISS, J.V. (2007): Biogeochemistry of aquifer systems. In: HURST, C.J., CRAWFORD, R.L., GARLAND, J.L., LIPSON, D.A., MILLS, A.L., STETZENBACH, L.D., Eds. (2007): Manual of Environmental Microbiology, 3rd edition. 843-859, ASM Press, Washington, US.

CUMMINGS, D.E., MARCH, A.W., BOSTICK, B., SPRING, S., CACCAVO JR., F., FENDORF, S., ROSENZWEIG, R.F. (2000): Evidence for microbial Fe(III) reduction in anoxic, mining-impacted lake sediments (Lake Coeur d'Alene, Idaho). Appl. Environ. Microbiol., 66, 154-162.

CUMMINGS, D.E., MAGNUSON, T.S. (2007): Microbial Fe(III) Reduction: Ecological and Physical Considerations. In: HURST, C.J., CRAWFORD, R.L., GARLAND, J.L., LIPSON, D.A., MILLS, A.L., STETZENBACH, L.D., Eds. (2007): Manual of Environmental Microbiology, 3rd edition. 1239-1248, ASM Press, Washington DC.

DAS, D., SAMANTA, G., MANDAL, B.K., CHOWDHURY, T.R., RANJAN, C., CHOWDHURY, P.P., BASU, G.K., CHAKRABORTI, D. (1996): Arsenic in groundwater in six districts of West Bengal, India. Environ. Geochem. Health, 18, 5-15.

DATES, G. (1994): Monitoring for phosphorus or how come they don't tell you this stuff in the manual? Vol. Mon., 6(1). Available from: http://water.epa.gov/type/rsl/ monitoring/upload/2004_10_13_monitoring_volunteer_newsletter_volmon06no1.pdf.

REFERENCES

DATTA, D.K., SUBRAMANIAN, V. (1997): Texture and mineralogy of sediments from the Ganges-Brahmaputra-Meghna river system in the Bengal Basin, Bangladesh and their environmental implications. Env. Geol., 30, 181-188.

DATTA, D.K., GUPTA, L.P., SUBRAMANIAN, V. (1999): Distribution of C, N and P in the sediments of the Ganges–Brahmaputra–Meghna river system in the Bengal basin. Org. Geochem., 30, 75-82.

DE LAETER, J.R., BÖHLKE, J.K., DE BIÈVRE, P., HIDAKA, H., TAYLOR, P.D.P. (2003): Atomic weights of the elements: Review 2000 (IUPAC Technical Report). Pure Appl. Chem., 75, 683–800.

DESCOLAS-GROS, C., SCHÖLZEL, C. (2007): Stable isotope ratios of carbon and nitrogen in pollen grains in order to characterize plant functional groups and photosynthetic pathway types. New Phytol., 176, 390-401.

DEUEL, L.E., SWOBODA, A.R. (1972): Arsenic mobility in a reduced environment. Soil Sci. Soc. Am. J., 36, 276–278.

DHAR, R.K., ZHENG, Y., STUTE, M., VAN GEEN, A., CHENG, Z., SHANEWAZ, M., SHAMSUDDUHA, M., HOQUE, M.A., RAHMAN, M.W., AHMED, K.M. (2008): Temporal variability of groundwater chemistry in shallow and deep aquifers of Araihazar, Bangladesh. J. Contam. Hydrol., 99, 97–111.

DIONEX Application note 133: Determination of Inorganic Anions in Drinking Water by Ion Chromatography. Available from: http://www.dionex.com/en-us/webdocs/4083-AN133_LPN1192-03.pdf.

DIXIT, S., HERING, J.G. (2003): Comparison of Arsenic(V) and Arsenic(III) Sorption onto Iron Oxide Minerals: Implications for Arsenic Mobility. Environ. Sci. Technol., 37, 4182-4189.

DIXIT, S., HERING, J.G. (2006): Sorption of Fe(II) and As(III) on goethite in single- and dual-sorbate systems. Chem. Geol., 228, 6-15.

DOLLASE, W.A. (1971): Refinement of the crystal structures of epidote, allanite and hancockit. Am. Min., 56, 447-464.

DORIOZ, J.M., PILLEBONE, E., FEHRI, A. (1989): Phosphorous dynamics in watersheds: role of trapping process in sediments. Water Res., 23, 147-158.

DRITS, V.A, ZVIAGINA, B.B., MC CARTY, D.K, SALYN, A.L. (2010): Factors responsible for crystal-chemical variations in the solid solutions from illite to aluminoceladonite and from glauconite to celadonite. Am. Min., 95, 348-361.

EICHE, E., NEUMANN, T., BERG, M., WEINMAN, B., VAN GEEN, A., NORRA, S., BERNER, Z., TRANG, P.T.K., VIET, P.H., STÜBEN, D. (2008): Geochemical processes underlying a sharp contrast in groundwater arsenic concentrations in a village on the Red River delta, Vietnam. Appl. Geochem., 23, 3143–3154.

EICHE, E. KRAMAR, U., BERG, M., BERNER, Z., NORRA, S., NEUMANN, T. (2010): Geochemical changes in individual sediment grains during sequential arsenic extractions. Water Res., 44, 5545-5555.

ENTENMANN, W. (2006): Baugrunderkundungen : planen - durchführen - überwachen – auswerten. Expert Verlag, Renningen.

EVANS, B.W., YANG, H. (1998): Fe-Mg order-disorder in tremolite-actinolite-ferro-actinolite at ambient and high temperature. Am. Min., 83, 458-475.

FAROOQ, S.H., CHANDRASEKHARAM, D., BERNER, Z., NORRA, S., STÜBEN, D. (2010): Influence of traditional agricultural practices on mobilization of arsenic from sediments to groundwater in Bengal delta. Water Res., 44, 5575–5588.

FENDORF, S., MICHAEL, H.A., VAN GEEN, A. (2010): Spatial and temporal variations of groundwater arsenic in south and Southeast Asia. Science, 328, 1123-1127.

FREEMAN, K.H., GOLDHABER, M.B. (2011): Future Directions in Geobiology and low-temperature Geochemistry. Elements, 7, 138-139.

FUJINO, Y., GUO, X., SHIRANR, K., LIU, J., WU, K., MIYATAKE, M., TANABE, K., KUSUDA, T., YOSHIMURA, T. (2006): Arsenic in Drinking Water and Peripheral Nerve Conduction Velocity among Residents of a Chronically Arsenic-affected Area in Inner Mongolia. Epidemiol., 16, 207-213.

GAT, J.R. (1996): Oxygen and hydrogen isotopes in the hydrologic cycle. Annu. Rev. Earth Planet. Sci., 24, 225–262.

GAULT, A.G., ISLAM, F.S., POLYA, D.A., CHARNOCK, J.M., BOOTHMAN, C., CHATTERJEE, D., LLOYD, J.R. (2005): Microcosm depth profiles of arsenic release in a shallow aquifer, West Bengal. Min. Mag., 69, 855-863.

GALLERT, C., WINTER J. (1997): Mesophilic and thermophilic anaerobic digestions of source-sorted organic waste: effect of ammonia on glucose degradation and methane production. Appl. Microbiol. Biotechnol., 48, 405-410.

REFERENCES

GAYE-HAAKE, B., LAHAJNAR, N., EMEIS, K.C., UNGER, D., RIXEN, T., SUTHOF, A., RAMASWAMY, V., SCHULZ, H., PAROPKARI, A.L., GUPTHA, M.V.S., ITTEKOT, V. (2005): Stable nitrogen isotopic ratios of sinking particles and sediments from the northern Indian Ocean. Marine Chem., 96, 243-255.

GIBBS-EGGAR, Z., JUDE, B., DOMINIK, J., LOIZEA, J.-L., OLDFIELD, F. (1999): Possible evidence for dissimilatory bacterial magnetite dominating the magnetic properties of recent lake sediments. Earth Plan. Sci. Lett., 168, 1-6.

GIBLING, M.R., BIRD, D.J. (1994): Late Carboniferous cyclothems and alluvial paleovalleys in the Sydney Basin, Nova Scotia. Geol. Soc. Am. Bull., 106, 105-117. In: MIAL, A.D., Ed. (1996): The Geology of Fluvial Deposits. Springer, Berlin, Heidelberg, New York.

GILHAM, R.W., STARR, R.C., MILLER, D.J. (1990): A device for in situ determination of Geochemical Transport Parameters 2. Biochemical reactions. Ground Water, 28, 858-862.

GLADNEY, E.S., ROELANDTS, I. (1990): Compilation of Elemental Concentration Data for USGS Geochemical Exploration Reference Materials GXR-1 to GXR-6. Geostd. Newslett., 14, 21-118.

GOH, K.H., LIM, T.T. (2004): Geochemistry of inorganic arsenic and selenium in a tropical soil: effect of reaction time, pH, and competitive anions on arsenic and selenium adsorption. Chemosph., 55, 849–859.

GOH, K.H., LIM, T.T. (2005): Arsenic fractionation in a fine soil fraction and influence of various anions on its mobility in the subsurface environment. Appl. Geochem., 20, 229–239.

GOODBRED JR., S.L., KUEHL, S.A., STECKLER, M.S., SARKA, M.H. (2003): Controls on facies distribution and stratigraphic preservation in the Ganges–Brahmaputra delta sequence. Sed. Geol., 155, 301–316.

GRAF, D.L. (1961): Crystallographic tables for the rhombohedral carbonates. Am. Min., 46, 1283-1316.

GOVINDARAJU, K. (1994): Compilation of working values and sample description for 383 geostandards. Geostand. Newslett., 18, 1-158.

GULENS, J., CHAMP, D.R., JACKSON, R.E. (1979): Influence of redox environments on the mobility of arsenic in groundwater. In: JENNE, E.A., Ed. (1979): Chemical Modelling in Aqueous Systems. ACS Symp. Ser., 93, 81–95.

GUO, H., STÜBEN, D., BERNER, Z. (2007): Removal of arsenic from aqueous solution by natural siderite and hematite. Appl. Geochem., 22, 1039–1051.

GUO, H., YANG, S., Tang, X., Li, Y., SHEN, Z. (2008): Groundwater geochemistry and its implications for arsenic mobilization in shallow aquifers of the Hetao Basin, Inner Mongolia. Sci. Tot. Env., 393, 131-144.

GUO, H., ZHANG, B., YANG, S., LI, Y., STÜBEN, D., NORRA, S., WANG, J. (2009): Role of colloidal particles for hydrogeochemistry in As-affected aquifers of the Hetao Basin, Inner Mongolia. Geochem. J., 43, 227-234.

HANSCOM, R. (1980): The structure of triclinic chloritoid and chloritoid polymorphism. Am. Min., 65, 534-539.

HANSEL, C.M., BENNER, S.G., NEISS, J., DOHNALKOVA, A., KUKKADAPU, R.K., FENDORF, S. (2003): Secondary mineralization pathways induced by dissimilatory iron reduction of ferrihydrite under advective flow. Geochim. Cosmochim. Acta, 67, 2977–2992.

HARVEY, C.F., SWARTZ, C.H., BADRUZZAMAN, A.B., KEON-BLUTE, N., YU, W., ALI, M.A. (2002): Arsenic mobility and groundwater extraction in Bangladesh. Science, 298, 1602-1606.

HARVEY, C.F., SWARTZ, C.H., BADRUZZAMAN, A.B.M., KEON-BLUTE, N., YU, W., ALI, M. A., JAY, J., BECKIE, R., NIEDAN, V., BRABANDER, D., OATES, P.M., ASHFAQUE, K.N., ISLAM, S., HEMOND, H.F., AHMED, M.F. (2003): Response to Comments on "Arsenic Mobility and Groundwater Extraction in Bangladesh". Science, 25, 584.

HARVEY, C.F., SWARTZ, C.H., BADRUZZAMAN, A.B.M., KEON-BLUTE, N., YU, W., ALI, M.A., JAY, J., BECKIE, R., NIEDAN, V., BRABANDER, D., OATES, P.M., ASHFAQUE, K.N., ISLAM, S., HEMOND, H.F., AHMED, M.F. (2005): Groundwater arsenic contamination on the Ganges Delta: biogeochemistry, hydrology, human perturbations, and human suffering on a large scale. C.R. Geosci., 337, 285-296.

HARVEY, C.F., ASHFAQUE, K.N., YU, W., BADRUZZAMAN, A.B.M., ALI, M.A., OATES, P.M., MICHAEL, H.A., NEUMANN, R.B., BECKIE, R., ISLAM, S., AHMED, M.F. (2006): Groundwater dynamics and arsenic contamination in Bangladesh. Chem. Geol., 228, 112-136.

HASSELHÖV, M., KAMMER, F. (2008): Iron Oxides as Geochemical Nanovectors for Metal Transport in Soil–River Systems. Elements, 4, 401–406.

HATTORI, K., TAKAHASI, Y., GUILLOT, S., JOHANSON, B. (2005): Occurrence of arsenic (V) in forearc mantle serpentinites based on X-ray absorption spectroscopy study. Geochim. Cosmochim. Acta, 69, 5585–5596.

HAYES, J.M. (1993): Factors controlling ^{13}C contents of sedimentary organic compounds: Principles and evidence. Marine Geol., 113, 111-125.

HERBEL, M., FENDORF, S. (2006): Biogeochemical processes controlling the speciation and transport of arsenic within iron coated sands. Chem. Geol., 228, 16–32.

HERING, J.G., KNEEBONE, P.E. (2002): Biogeochemical Controls on Arsenic Occurrence and Mobility in Water Supplies. In: FRANKENBERGER JR., W.T., Ed. (2002): Environmental Chemistry of Arsenic. 155-182, Marcel Dekker, Inc., New York.

HÉRY, M., GAULT, A.G., ROWLAND, H.A.L., LEAR, G., POLYA, D.A., LLOYD, J.R. (2008): Molecular and cultivation-dependent analysis of metal-reducing bacteria implicated in arsenic mobilisation in south-east Asian aquifers. Appl. Geochem., 23, 3215-3223.

HÉRY, M., VAN DONGEN, B.E., GILL, F., MONDAL, D., VAUGHAN, D.J., PANCOST, R.D., POLYA, D.A., LLOYD, J.R. (2010): Arsenic release and attenuation in low organic carbon aquifer sediments from West Bengal. Geomicrobiol., 8, 155-168.

HEROY, D.C., KUEHL, S.A., GOODBRED JR., S.L. (2003): Mineralogy of the Ganges and Brahmaputra Rivers: implications for river switching and Late Quaternary climate change. Sed. Geol. 155, 343–359.

HINDMARSH, J.T., CURDY, R.F. (1986): Clinical and environmental aspects of arsenic toxicity. CRC Crit. Rev. Clin. Lab. Sci., 23, 315–347. In: WHO (2003): Arsenic in Drinking-water - Background document for development of WHO Guidelines for Drinking-water Quality. WHO/SDE/WSH/03.04/75. Available from http://www.who.int/ water_sanitation_health/dwq/ chemicals/arsenic.pdf.

HOEFS, J. (2009): Stable isotope geochemistry, 6th edition. Springer, Berlin, Heidelberg, New York.

HOPENHAYN, C. (2006): Arsenic in Drinking Water: Impact on Human Health. Elements, 2, 103-107.

HORNIBROOK, E.C.R., LONGSTAFFE, F.J., FYFE, W.S., BLOOM, Y. (2000): Carbon-isotope ratios and carbon, nitrogen and sulfur abundances inflora and soil organic matter from a temperate-zone bog and marsh. Geochem. J., 34, 237-245.

236

IAEA (1993): Reference and intercomparison materials for stable isotopes of light elements Proceedings of a consultants meeting held in Vienna, 1-3 December 1993. Available from: http://www-pub.iaea.org/MTCD/publications/PDF/te_825_prn.pdf.

INDIAN STANDARD SPECIFICATIONS FOR DRINKING WATER (1992): IS 10500. Available from: http://hppcb.gov.in/eiasorang/spec.pdf.

INSKEEP, W.P., MC DERMOTT, T.R., FENDORF, S. (2002): Arsenic (V)/(III) Cycling in Soils and Natural Waters: Chemical and Microbiological Processes. In: FRANKENBERGER JR., W.T., Ed. (2002): Environmental Chemistry of Arsenic. 183-215, Marcel Dekker Inc., New York.

IPCS (2001): Environmental Health Criteria 224: Arsenic and arsenic compounds, second edition. World Health Organization, Geneva, Switzerland. Available from: http://www.bvsde.ops-oms.org/bvsacg/e/cdcagua/guias/b.parametos/4.BasTox/IPCS/008.arsenic.pdf.

ISLAM, F.S., GAULT, A.G., BOOTHMAN, C., POLYA, D.A., CHARNOCK, J.M., CHATTERJEE, D., LLOYD, J.R. (2004): Role of metal-reducing bacteria in arsenic release from Bengal delta sediments. Nature, 430, 68-71.

ISLAM, F.S., PEDERICK, R.L., GAULT, A.G., POLYA, D.A. (2005): Interactions between the Fe(III)-reducing bacterium *Geobacter sulfurreducens* and arsenate, and capture of the metalloid by biogenic Fe(II). Appl. Environ. Microbiol., 71, 8642-8648.

ISLAM, F.S. (2008): Microbial controls on the geochemical behaviour of arsenic in groundwater systems. In: AHUJA, S., Ed. (2008): Arsenic contamination of groundwater: Mechanism, Analysis, and Remediation. 51-83, John Wiley & Sons, Hoboken, New Jersey.

ISLAM, M.S., TOOLEY, M.J. (1999): Coastal and sea-level changes during the Holocene in Bangladesh. Quat. Int., 55, 61-75.

ITAI, T., TAKAHASHI, Y., SEDDIQUE, A.A., MARUOKA, T., MITAMURA, M. (2010): Variations in the redox state of As and Fe measured by X-ray absorption spectroscopy in aquifers of Bangladesh and their effect on As adsorption. Appl. Geochem., 25, 34-47.

JACKSON, C.R., LANGNER, H.W., DONAHOE-CHRISTANSEN, J., INSKEEP, W.P., MC DERMOTT, T.R. (2001): Molecular analysis of microbial community structure in an arsenite-oxidizing acidic thermal spring. Environ. Microbiol., 3, 532–542.

JICA (2003): Annual Report 2003. Available from: http://www.jica.go.jp/english/publications/reports/ annual/2003/index.html.

REFERENCES

JOHNSTON, R.B., BERG, M., JOHNSON, C.A., TILLEY, E., HERING, J.G. (2011): Water and sanitation in developing countries; geochemical aspects of quality and treatment. Elements, 7, 163-168.

JÖNSSON, J., SHERMAN, D.M. (2008): Sorption of As(III) and As(V) to siderite, green rust (fougerite) and magnetite: Implications for arsenic release in anoxic groundwaters. Chem. Geol., 225, 173-181.

JURGENS, B.C., MC MAHON, P.B., CHAPELLE, F.H., EBERTS, S. (2009): An Excel® workbook for identifying redox processes in ground water: U.S. Geological Survey Open-File Report, 2009-1004. Available from: http://pubs.usgs.gov/of/2009/b1004/ofr_2009-1004.pdf.

KAPAJ, S., PETERSON, H., LIBER, K., BHATTACHARYA, P. (2006): Human Health Effects From Chronic Arsenic Poisoning - A Review. J. Environ. Sci. Health Pt. A, 41, 2399-2428.

KAPPLER, A., BENZ, M., SCHINK, B., BRUNE, A. (2004): Electron shuttling via humic acids in microbial iron(III) reduction in a freshwater sediment. FEMS Microbiol. Ecol., 47, 85-92.

KAPPLER, A. (2011): Personal communication at IDRAS workshop in Hanoi.

KEON, E., SWARTZ, C.H., BRABANDER, D.J., HARVEY, C., HEMOND, H.F. (2001): Validation of an arsenic sequential extraction method for evaluating mobility in sediments. Environ. Sci. Technol., 35, 2778–2784.

KIEFT, T.L., PHELPS, T.J., FREDRICKSON, J.K. (2007): Drilling, Coring, and Sampling Subsurface Environments. In: HURST, C.J., CRAWFORD, R.L., GARLAND, J.L., LIPSON, D.A., MILLS, A.L., STETZENBACH, L.D., Eds. (2007): Manual of Environmental Microbiology, 3rd edition. 799-817, ASM Press, Washington.

KINNIBURGH, D.G., SMEDLEY, P.L., Eds. (2001): Arsenic contamination of groundwater in Bangladesh, Vol. 1: Summary. Bangladesh Department for Public Health Engineering and British Geological Survey, Keyworth. Available from: http://www.bgs.ac.uk/arsenic.

KIRK, M.F., RODEN, E.E., CROSSEY, L.J., BREALEY, A.J., SPILDE, M.N. (2010): Experimental analysis of arsenic precipitation during microbial sulphate and iron reduction in model aquifer sediment reactors. Geochim. Cosmochim. Acta, 74, 2538-2555.

KNOWELS, F.C., BENSON, A.A. (1983): The biochemistry of arsenic. Trends in Biochem. Sci., 8, 178-180.

KOCAR, B.D., HERBEL, M.J., TUFANO, K.J., FENDORF, S. (2006): Contrasting effects of dissimilatory iron(III) and arsenic(V) reduction on arsenic retention and transport. Environ. Sci. Technol., 40, 6715-6721.

KOCAR, B.D., POLIOZOTTO, M.L., BEBBER, S.G., YING, S.C., UNG, M., OUCH, K., SAMRETH, S., SUY, B., PHAN, K., SAMPSON, M., FENDORF, S. (2008): Integrated biogeochemical and hydrologic processes driving arsenic release from shallow sediments to groundwaters of the Mekong delta. Appl. Geochem., 23, 3059–3071.

KOCAR, B.D., FENDORF, S. (2009): Thermodynamic Constraints on Reductive Reactions Influencing the Biogeochemistry of Arsenic in Soils and Sediments. Environ. Sci. Technol., 43, 4871-4877.

KONHAUSER, K.O. (2007): Introduction to Geomicrobiology. Blackwell Science Ltd., Oxford.

KONHAUSER, K.O., KAPPLER, A., RODEN, E.E. (2011): Iron in microbial metabolisms. Elements, 7, 89-93.

LAMB, A.L., WILSON, G.P., LENG, M.L. (2006): A review of coastal palaeoclimate and relative sea-level reconstructions using $d^{13}C$ and C/N ratios in organic material. Earth Sci. Revs., 75, 29-57.

LAMBECK, K., ESAT, T.M., POTTER, E.-K. (2002): Links between climate and sea levels for the past three million years. Nature, 219, 199-206.

LANGMUIR, D. (1997): Aqueous Environmental Geochemistry. Prentice Hall, New Jersey. In: ZHU, C., SCHWARTZ, W. (2011): Hydrogeochemical Processes and Controls on Water Quality and Water Management. Elements, 7, 171.

LAPHAM, W.W., WILDE, F.D., KOTERBA, M.T., Eds. (1997): Guidelines and standard procedures for studies of ground-water quality: selection and installation of wells, and supporting documentation. U.S. Geological Survey Water Resources Investigation Report, U.S. Geological Survey: Reston, Virginia. Available from: http://water.usgs.gov/owq/pubs/wri/wri964233/wri964233.pdf.

LAWSON, M., BALLENTINE, C.J., POLYA, D.A., BOYCE, A.J., MONDAL, D., CHATTERJEE, D., MAJUMDER, S., BISWAS, A. (2008): The geochemical and isotopic composition of ground waters in West Bengal: tracing ground-surface water interaction and its role in arsenic release. Min. Mag., 72, 441-444.

REFERENCES

LE, X.C. (2002): Arsenic speciation in the environment and humans. In: FRANKENBERGER JR., W.T., Ed. (2002): Environmental Chemistry of Arsenic. 95-116, Marcel Dekker Inc., New York.

LEAR, G., SONG, B., GAULT, A.G., POLYA, D.A., LLOYD, J.R. (2007): Molecular analysis of arsenate-reducing bacteria within Cambodian sediments following amendment with acetate. Appl. Environ. Microbiol., 73, 1041-1048.

LEVIEN, L., PREWITT, C.T., WEIDNER, D.J. (1980): Structure and elastic properties of quartz at pressure. Am. Min., 65, 920-930.

LEWANDOWSKI, J., NÜTZMANN, G. (2010): Nutrient retention and release in a floodplain's aquifer and in the hyporheic zone of a lowland river. Ecol. Eng., 36, 1156–1166.

LIN, N.-F., TANG, J., BIAN, J.-M. (2004): Geochemical environment and health problems in China. Environ. Geochem. Health, 26, 81–88.

LIN, T.-H., HUANG, Y.-L., WANG, M.-Y. (1998): Arsenic species in drinking water, hair, fingernails, and urine in patients with blackfoot disease. J. Tox. Environ. Health Pt. A, 53, 85-93.

LINDSAY, J.F., HOLIDAY, D.W., HULBERT, A.G. (1991): Sequence stratigraphy and the evolution of the Ganges–Brahmaputra complex. Am. Assoc. Petrol. Geol. Bull., 75, 1233–1254. In: MUKHERJEE, A., FRYA, A.E., THOMAS, W.A. (2009): Geologic, geomorphic and hydrologic framework and evolution of the Bengal basin, India and Bangladesh. J. Asian Earth Sci., 34, 227–244.

LLOYD, J.R., OREMLAND, R.S. (2006): Microbial Transformations of Arsenic in the Environment: From Soda Lakes to Aquifers. Elements, 2, 85-90.

LOVLEY, D.R., PHILLIPS, E.J.P. (1986): Organic matter mineralization with reduction of ferric iron in anaerobic sediments. Appl. Environ. Microbiol., 51, 683-689.

LOVLEY, D.R., PHILLIPS, E.J.P. (1987): Rapid Assay for Microbially Reducible Ferric Iron in Aquatic Sediments. Appl. Environ. Microbiol., 53, 1536-1540.

LOVLEY, D.R., PHILLIPS, E.J.P. (1989): Requirement for a microbial consortium to completely oxidize glucose in Fe(III)-reducing sediments. Appl. Environ. Microbiol., 55, 3234-3236.

LOVLEY, D.R. (1993): Dissimilatory Metal Reduction. Annu. Rev. Microbiol., 47, 263-290.

LOVLEY, D.R., CHAPELLE, F.H., WOODWARD, J.C. (1994): Use of dissolved H_2 concentrations to determine distribution of microbially catalyzed redox reactions in anoxic groundwater. Environ. Sci. Technol., 28, 1205-1210.

LOVLEY, D.R. (1995): Microbial reduction of iron, manganese, and other metals. Adv. Agron., 54, 175–231.

LOVLEY, D.R., COATES, J.D., BLUNT-HARRIS, E.L., PHILLIPS, E.J.P., WOODWARD, J.C. (1996): Humic substances as electron acceptors for microbial respiration. Nature, 382, 445-448.

LOVLEY, D.R. (1997): Microbial Fe(III) reduction in subsurface environments. FEMS Microbiol. Rev., 20, 305-315.

LOVLEY, D.R., HOLMES, D.E., NEVIN, K.P. (2004): Dissimilatory Fe(III) and Mn(IV) reduction. Adv. Microb. Physiol., 49, 219-286.

LU, P., ZHU, C. (2011): Arsenic Eh–pH diagrams at 25°C and 1bar. Environ. Earth Sci., 62, 1673–1683.

MOHAN, D., PITTMAN JR., C.U. (2007): Arsenic removal from water/wastewater using adsorbents – A critical review. J. Hazard. Mat., 142, 1-53.

MAILLOUX, B.J., ALEXANDROVA, E., KEIMOWITZ, A.R., WOVKULICH, K., FREYER, G.A., HERRON, M., STOLZ, J.F., KENNA, T.C., PICHLER, T., POLIZZOTTO, M.L., DONG, H., BISHOP, M., KNAPPET, P.S.K. (2009): Microbial Mineral Weathering for Nutrient Acquisition Releases Arsenic. Appl. Env. Microbiol., 75, 2558–2565.

MALASARN, D., SALTIKOV, C.W., CAMPBELL, K.M., SANTINI, J.M., HERING, J.G., NEWMAN, D.K. (2004): *arrA* Is a Reliable Marker for As(V) Respiration. Science, 306, 455.

MARCHAND, C., DISNAR, J.R., LALLIER-VERGÉS, E., LOTTIER, N. (2005): Early diagenesis of carbohydrates and lignin in mangrove sediments subject to variable redox conditions (French Guiana). Geochem. Cosmochem. Acta, 69, 131-142.

MATSCHULLAT, J. (2000): Arsenic in the geosphere- a review. Sci. Tot. Environ., 249, 297-312.

MC ARTHUR, J.M., RAVENSCROFT, P., SAFIULLA, S., THIRLWALL, F. (2001): Arsenic in groundwater: Testing pollution mechanisms for sedimentary aquifers in Bangladesh. Water Resour. Res., 37, 109-117.

REFERENCES

MC ARTHUR, J.M., BANERJEE, D.M., HUDSON-EDWARDS, K.A., MISHRA, R., PUROHIT, R., RAVENSCROFT, P., CRONIN, A., HOWARTH, R.J., CHATTERJEE, A., TALUKDER, T., LOWRY, D., HOUGHTON, S., CHADHA, D.K. (2004): Natural organic matter in sedimentary basins and its relation to arsenic in anoxic ground water: the example of West Bengal and its worldwide implications. Appl. Geochem., 19, 1255-1293.

MC ARTHUR, J.M., RAVENSCROFT, P., BANERJEE, D.M., MILSOM, J., HUDSON-EDWARDS, K.A., SENGUPTA, S., BRISTOW, C., SARKAR, A., TONKIN, S., PUROHIT, R. (2008): How paleosols influence groundwater flow and arsenic pollution: A model from the Bengal Basin and its worldwide implication. Water Resour. Res., 44 (W11411), 1-30.

MC ARTHUR, J.M., BANERJEE, D.M., SENGUPTA, S., RAVENSCROFT, P., KLUMP, S., SARKAR, A., BISCH, D., KIPFER, R. (2010): Migration of As, and ^{3}H/^{3}He ages in groundwater from West Bengal: Implications for monitoring. Water Res., 44, 4171-4185.

MC ARTHUR, J.M., NATH, B., BANERJEE, D.M., PUROHIT, R., GRASSINEAU, N. (2011): Palaeosol Control on Groundwater Flow and Pollutant Distribution: The Example of Arsenic. Environ. Sci. Technol., 45, 1376–1383.

MC ARTHUR, J.M., SIKDAR, P.K., NATH, B., GRASSINEAU, N., MARSHALL, J.D., BANERJEE, D.M. (2012): Sedimentological Control on Mn, and Other Trace Elements. In: Groundwater of the Bengal Delta. Environ. Sci. Technol., 46, 669–676.

MC MAHON, P.B., CHAPELLE, F.H. (2008): Redox processes and water quality of selected principal aquifer systems. Ground Water, 46, 259-271.

MELIKER, J.R., NRIAGU, J.O. (2007): Arsenic in drinking water and bladder cancer: review of epidemiological evidence. In: BHATTACHARYA, P., MUKHERJEE, A.B., BUNDSCHUH, J., ZEVENHOVEN, R., LOEPPERT, R.H., Eds. (2007): Arsenic in soil and groundwater environment: Biogeochemical Interactions, Health Effects and Remediation (Trace Metals and Other Contaminants in the Environment). 551-586, Elsevier, Amsterdam.

MENG, X., KORFIATIS, G.P., JING, C., CHRISTODOULATOS, C. (2001): Redox transformations of arsenic and iron in water treatment sludge during aging and TCLP extraction. Environ. Sci. Technol., 35, 3476-3481.

MERKEL, B.J., PLANER-FRIEDRICH, B. (2008): Grundwasserchemie – Praxisorientierter Leitfaden zur numerischen Modellierung von Beschaffenheit, Kontamination und Sanierung aquatischer Systeme. 2. Auflage. Springer, Heidelberg.

MÉTRAL, J., CHARLET, L., BUREAU, S., MALLIK, S.B., CHAKRABORTY, S., AHMED, K.M., RAHMAN, M.W., CHENG, Z., VAN GEEN, A. (2008): Comparison of dissolved and particulate arsenic distributions in shallow aquifers of Chakdaha, India, and Araihazar, Bangladesh. Geochem. Trans., 9 (1), DOI:10.1186/1467-4866-9-1.

MEYERS, P.A. (1994): Preservation of elemental and isotopic source identification of sedimentary organic matter. Isotope Geosci., 289-302.

MIAL, A.D. (1996): The Geology of Fluvial Deposits. Springer, Berlin, Heidelberg, New York.

MICHAEL, H.A., VOSS, C.I. (2009a): Estimation of regional-scale groundwater flow properties in the Bengal Basin of India and Bangladesh. Hydrogeol. J., 17, 1329-1346.

MICHAEL, H.A., VOSS, C.I. (2009b): Controls on groundwater flow in the Bengal Basin of India and Bangladesh: Regional modeling analysis. Hydrogeol. J., 17, 1561–1577.

MLADENOV, N., ZJENG, Y., MILLER, M.P., NEMERGUT, D.R., LEGG, T., SIMONE, B., HAGEMAN, C., RAHMAN, M.M., AHMED, K.M., MC KNIGHT, D.M. (2010): Dissolved organic matter sources and consequences for Iron and Arsenic Mobilization in Bangladesh Aquifers. Environ. Sci. Technol., 44, 123–128.

MOHAN, D., PITMAN, C.U. Jr. (2007): Arsenic removal from water/wastewater using adsorbents – A critical review. J. Hazard. Mat., 142, 1-53.

MOORE, D.M., REYNOLDS, R.C. (1989): X-ray diffraction and the identification and analysis of clay minerals. Oxford University Press, Oxford, New York.

MORIN, G., CALAS, G. (2006): Arsenic in Soils, Mine Tailings, and Former Industrial Sites. Elements, 2, 97-101.

MUKHERJEE, A., FRYA, A.E., HOWELL, P.D. (2007): Regional hydrostratigraphy and groundwater flow modelling in the arsenic-affected areas of the western Bengal basin, West Bengal, India. Hydrogeol. J., 15, 1397-1418.

MUKHERJEE, A., VON BRÖEMSSEN, M., SCANLON, B.R., BHATTARACHARYA, P., FRYAR, A.E., HASAN, M.A., AHMED, K.M., JACKS, G., CHATTERJEE, D., SRACEK, O. (2008): Hydrogeochemical comparison and effects of overlapping redoxzones on groundwater arsenic near the western (Bhagirathisub-basin, India) and eastern (Meghnasub-basin Bangladesh) of the Bengal basin. J. Cont. Hydrol., 99, 31-48.

REFERENCES

MUKHERJEE, A., FRYAR, A.E., THOMAS, W.A. (2009): Geologic, geomorphic and hydrologic framework and evolution of the Bengal basin, India and Bangladesh. J. Asian Earth Sci., 34, 227–244.

MUKHERJEE, A., FRYAR, A.E., SCANLON, B.R., BHATTARACHARYA, P., BHATTARACHARYA, A. (2011): Elevated arsenic in deeper groundwater of the western Bengal basin, India: Extent and controls from regional to local scale. Appl. Geochem., 26, 600–613.

MÜLLER, K., CIMINELLI, V.S.T., DANTAS, M.S.D., WILLSCHER, S. (2010): A comparative study of As(III) and As(V) in aqueous solutions and adsorbed on ironoxy-hydroxides by Raman spectroscopy. Water Res., 44, 5660-5672.

MUZUKA, A.N.N., SHUNULA, J.P. (2006): Stable isotope compositions of organic carbon and nitrogen of two mangrove stands along the Tanzanian coastal zone. Estuarine Coastal Shelf Sci., 66, 447-458.

NATH, B., STÜBEN, D., BASU-MALLIK, S., CHATTERJEE, D., CHARLET, L. (2008): Mobility of arsenic in West Bengal aquifers conducting low and high groundwater arsenic. Part I: Comparative hydrochemical and hydrogeological characteristics. Appl. Geochem., 23, 977–995.

NEIDHARDT, H., NORRA, S., TANG, X., GUO, H., STÜBEN, D. (2012a): Impact of irrigation with high arsenic burdened groundwater on the soil-plant system: Results from a case study in the Inner Mongolia, China. Environ. Pol., 163, 8-13.

NEIDHARDT, H., BERNER, Z., FREIKOWSKI, D., BISWAS, A., MAJUMDER, S., WINTER, J., GALLERT, C., CHATTERJEE, D., NORRA, S., (2012b): On the role of geomicrobiological processes in the mobilization of Mn and As in the Bengal Delta Plain. Submitted to Environ. Sci. & Technol.

NEIDHARDT, H., BERNER, Z., FREIKOWSKI, D., BISWAS, A., WINTER, J., CHATTERJEE, D., NORRA, S., (2012c): Influences of groundwater extraction on the distribution of dissolved As in shallow aquifers of West Bengal, India. Submitted to J. Hazard. Mat.

NEUMANN, R.B., ASHFAQUE, K.N., BADRUZZAMAN, A.B.M., ALI, M.A., SHOEMAKER, J.K., HARVEY, C.H. (2009): Anthropogenic influences on groundwater arsenic concentrations in Bangladesh. Nature Geosci., 3, 46-52.

NEWMAN, D.K., KENNEDY, E.K., COATES, J.D., AHMANN, D., ELLIS, D.J., LOVLEY, D.R., MOREL, F.M.M. (1998): Dissimilatory arsenate and sulphate reduction in *Desulfotomaculum auripigmentum*. Archiv. Microbiol., 165, 380–388.

NICKSON, R., MC ARTHUR, J.M., BURGES, W.G., AHMED, K.M., RAVENSCROFT, P., RAHMAN, M. (1998): Arsenic poisoning of Bangladesh groundwater. Nature, 395, 338.

NICKSON, R., MC ARTHUR, J.M., RAVENSCROFT, P., BURGES, W.G., AHMED, K.M. (2000): Mechanism of arsenic release to groundwater, Bangladesh and West Bengal. Appl. Geochem.,15, 403–413.

NORRA, S., BERNER, Z.A., AGARWALA, P., WAGNER, F., CHANDRASEKHARAM, D., STÜBEN, D. (2005): Impact of irrigation with As rich groundwater on soil and crops: a geochemical case study in West Bengal Delta Plain, India. Appl. Geochem., 20, 1890-1906.

NORRMAN, J., SPARRENBOM, C.J., BERG, M., NHAN, D.D., NHAN, P.Q., ROSQVIST, H., JACKS, G., SIGYARDSSON, E., BARIC, D., MORESKOG, J., HARMS-RINGDAHL, P., VAN HOAN, N. (2008): Arsenic mobilisation in a new well field for drinking water production along the Red River, NamDu, Hanoi. Appl. Geochem., 23, 3127–3142.

NOVAK, G.A., GIBBS, G.V. (1971): The crystal chemistry of the silicate garnets. Am. Min., 56, 791-825.

NRIAGU, J.O., Ed. (1994): Arsenic in the Environment, Part II: Human Health and Ecosystem Effects. John Wiley & Sons, N.Y.

NRIAGU, J.O., BHATTACHARYA, P., MUKHERJEE, A.B., BUNDSCHUH, J., ZEVENHOVEN, R., LOEPPERT, R.H. (2007): Arsenic in soil and groundwater: an overview. In: BHATTACHARYA, P., MUKHERJEE, A.B., BUNDSCHUH, J., ZEVENHOVEN, R., LOEPPERT, R.H., Eds. (2007): Arsenic in soil and groundwater environment: Biogeochemical Interactions, Health Effects and Remediation (Trace Metals and Other Contaminants in the Environment). 3-63, Elsevier, Amsterdam.

O'DAY, P.A., VLASSOPOULOS, D., ROOT, R., RIVERA, N. (2004): The influence of sulphur and iron on dissolved arsenic concentrations in the shallow subsurface under changing redox conditions. PNAS, 101, 13703-13708.

O'DAY, P.A. (2006): Chemistry and Mineralogy of Arsenic. Elements, 2, 77-83.

O'LOUGHLIN, E.J., GORSKI, C.A., SCHERER, M.M., BOYANOV, M.I., KEMNER, K.M. (2010): Effects of oxyanions, natural organic matter, and bacterial cell numbers on the bioreduction of lepidocrocite (γ-FeOOH) and the formation of secondary mineralization products. Environ. Sci. Technol., 44, 4570-5476.

REFERENCES

ONA-NGUEMA, G., MORIN, G., JUILLOT, F., CALAS, G., BROWN JR., G.E. (2005): EXAFS Analysis of Arsenite Adsorption onto Two-Line Ferrihydrite, Hematite, Goethite, and Lepidocrocite. Environ. Sci. Technol., 39, 9147-9155.

OREMLAND, R.S., STOLZ, J.F. (2005): Arsenic, microbes and contaminated aquifers. Trends Microbiol., 13, 45-49.

PAGE, W.J., HUYER, M. (1984): Derepression of the *Azotovacter vinelandii* siderophore system, using iron-containing minerals to limit iron repletion. In: KONHAUSER, K.O., KAPPLER, A., RODEN, E.E. (2011): Iron in microbial metabolisms. Elements, 7, 89-93.

PAL, T., MUKHERJEE, P.K., SENGUPTA, S., BATTACHARYYA, A.K., SHOME, S. (2002): Arsenic pollution in groundwater of West Bengal, India – an insight into the problem by subsurface sediment analysis. Gondwana Res. 5, 501-512.

PANSU, M., GAUTHEROU, J. (2006): Handbook of soil analysis. Mineralogical, organic and inorganic methods. 180-191, Springer, Berlin, Heidelberg.

PARKHURST, D.L., APPELO, C.A.J. (1999): User's guide to PHREEQC (version 2) - a computer program for speciation, reaction-path, 1D-transport, and inverse geochemical calculations, US Geological Survey Water Resources Investigation Report 99, 99-4259, U.S. Geological Survey, Reston, Virginia. Available from: http://wwwbrr.cr.usgs.gov/projects/GWC_coupled/phreeqc/.

PEDERSEN, H.D., POSTMA, D., JAKOBSEN, R., LARSON, O. (2005): Fast transformation of iron oxyhydroxides by the catalytic action of aqueous Fe(II). Geochim. Cosmochim. Acta, 69, 3967-3977.

PEDERSEN, H.D., POSTMA, D., JAKOBSEN, R. (2006): Release of arsenic associated with the reduction and transformation of iron oxides. Geochim. Cosmochim. Acta, 70, 4116-4129.

PÉREZ, S. (1995): The structure of sucrose in the crystal and in solution. In: MATHLOUTHI, M., REISER, P., Eds. (1995): Sucrose: Properties and applications. 11-32, Chapman & Hall, Suffolk.

PHILIPS, T.L., LOVELESS, J.K., BAILEY, S.W. (1980): Cr^{3+} coordination in chlorites: a structural study of ten chromian chlorites. Am. Min., 65, 112-122.

POLIOZOTTO, M.L., KOCAR, B.D., BENNER, S.G., SAMPSON, M., FENDORF, S. (2008): Near-surface wetland sediments as a source of arsenic release to groundwater in Asia. Nature, 454, 505-508.

POLYA, D., CHARLET, L. (2009): Rising arsenic risk? Nat. Geosci., 2, 383-384.

POMROY, C., CHARBONNEAU, S.M., MC CULLOUGH, R.S., TAM, G.K.H. (1980): Human retention studies with ^{74}As. Toxicol. Appl. Pharmacol., 53, 550–556.

POSTMA, D., LARSEN, F., HUE, N.T.M., DUC, M.T., VIET, P.H., NHAN, P.Q., JESSEN, S. (2007): Arsenic in groundwater of the Red River floodplain, Vietnam: Controlling geochemical processes and reactive transport modeling. Geochim. Cosmochim. Acta, 71, 5054-5071.

POSTMA, D., JESSEN, S., HUE, N.T.H., DUC, M.T., KOCH, C.B., VIET, P.H., NHAN, P.Q., LARSEN, F. (2010): Mobilization of arsenic and iron from Red River floodplain sediments, Vietnam. Geochim. Cosmochim. Acta, 74, 3367–3381.

QUICKSALL, A.N., BOSTICK, B.C., SAMPSON, M.L. (2008): Linking organic matter deposition and iron mineral transformations to groundwater arsenic levels in the Mekong delta, Cambodia. Appl. Geochem., 23, 3088-3098.

RADLOFF, K.A., CHENG, Z., RAHMAN, M.W., AHMED, K.M., MAILLOUX, B.J., JUHL, A.R. (2007): Mobilization of arsenic during one-year incubations of grey aquifer sands from Araihazar, Bangladesh. Environ. Sci. Technol., 41, 3639-3645.

RADLOFF, K.A., MANNING, A.R., MAILLOUX, B., ZHENG, Y., RAHMAN, M., HUQ, M.R., AHMED, K.M., VAN GEEN, A. (2008): Considerations for conducting incubations to study the mechanisms of As release in reducing groundwater aquifers. Appl. Geochem., 23, 3224-3235.

RAHMAN, A., VAHTER, M., EKSTRÖM, E.C. (2007): Association of arsenic exposure during pregnancy with fetal loss and infant death: a cohort study in Bangladesh. Am. J. Epidemiol., 165, 1389-1396.

RAHMAN, M., TONDEL, M., AHMAD, S.A., AXELSON, O. (1998): Diabetes mellitus associated with arsenic exposure in Bangladesh. Am. J. Epidemiol., 148 (2), 198-203.

RAISWELL, R. (2011): Iron Transport from the Continents to the Open Ocean: The Aging-Rejuvenation Cycle. Elements, 7, 101-106.

RAVENSCROFT, P., MC ARTHUR, J.M., HOQUE, B.A. (2001): Geochemical and Palaeohydrological Controls on Pollution of Groundwater by Arsenic. In: CHAPPEL, W.R., ABERNATHY, C.O., CALDERON, R., Eds. (2001): Arsenic Exposure and Health Effects IV. 53-78, Elsevier, Oxford.

REDDY, K.R., KADLEC, R.H., FLAIG, E., GALE, P.M. (1999): Phosphorous retention in streams and wetlands: A review. Crit. Rev. Environ Sci. Technol., 29, 83-146.

REFERENCES

RICHARDSON, S.M., RICHARDSON, J.W. (1982): Crystal structure of a pink muscovite from Archer's Post, Kenya: Implications for reverse pleochroism in dioctahedral micas. Am. Min., 67, 69-75.

RODELLI, M.R., GEARING, J.N., GEARING, P.J., MARSHALL, N., SASEKUMAR, A. (1984): Stable isotope ratio as a tracer of mangrove carbon in Malaysian ecosystems. Oecologica, 61, 326–333.

RODEN, E.E. (2003): Fe(III) Oxide Reactivity Toward Biological versus Chemical Reduction. Env. Sci. Technol., 37, 1319-1324.

RODEN, E.E., KAPPLER, A., BAUER, I., PAUL, A., STOESSER, R., KONISHI, H., XU, H. (2010): Extracellular electron transfer through microbial reduction of solid-phase humic substances. Nature Geosci., 3, 417-421.

ROWLAND, H.A.L., GAULT, A.G., CHARNOCK, J.M., POLYA, D.A. (2005): Preservation and XANES determination of the oxidation state of solid-phase arsenic in shallow sedimentary aquifers in Bengal and Cambodia. Mineral. Mag., 69, 825-839.

ROWLAND, H.A.L., POLYA, D.A., LLOYD, J.R., PANCOST, R.D. (2006): Characterisation of organic matter in a shallow, reducing, arsenic-rich aquifer, West Bengal. Org. Geochem., 37, 1101–1114.

ROWLAND, H.A.L., PEDERICK, R.L., POLYA, D.A., PANCOST, R.D., VAN DONGEN, B.E., GAULT, A.G., VAUGHAN, D.J., BRYANT, C., ANDERSON, B., LLOYD, J.R. (2007): The control of organic matter on microbially mediated iron reduction and arsenic release in shallow alluvial aquifers, Cambodia. Geobiol., 5, 281–292.

ROWLAND, H.A.L., BOOTHMAN, C., PANCOST, R., GAULT, A.G., POLYA, D.A, LLOYD, J.R. (2009): The Role of Indigenous Microorganisms in the Biodegradation of Naturally Occurring Petroleum, the Reduction of Iron, and the Mobilization of Arsenite from West Bengal Aquifer Sediments. J. Environ. Qual., 38, 1598-1607.

SALMASSI, T.M., VENKATESWAREN, K., SATOMI, M., NEALSON, K.H., NEWMAN, D.K., HERING, J.G. (2002): Oxidation of Arsenite by *Agrobacterium albertimagni*, AOL15, sp. nov., Isolated from Hot Creek, California. Geomicrobiol. J., 19, 53-66.

SANDELL, E.B. (1959): Colorimetric determination of traces of metals; 3rd edition. Interscience Publishers Inc., New York.

SARKA, A., SENGUPTA, S., MC ARTHUR, J.M., RAVENSCROFT, P., BERA, M.K., BHUSHAN, R., SAMANTA, A., AGRAWAL, S. (2009): Evolution of Ganges–Brahmaputra western delta plain: clues from sedimentology and carbon isotopes. Quat. Sci. Rev., 28, 2564-2581.

SAUNDERS, J.A., LEE, M.-K., SHAMSUDDUHA, M., DHAKAL, P., UDDIN, A., CHOWDURY, M.T., AHMED, K.M. (2008): Geochemistry and mineralogy of arsenic in (natural) anaerobic groundwaters. Appl. Geochem., 23, 3205-3214.

SCHOELL, M., FABER, E., COLEMAN, M.L. (1983): Carbon and hydrogen isotopic compositions of the NBS 22 and NBS 21 stable isotope reference materials: an inter-laboratory comparison. Org. Geochem., 5, 3–6.

SCHUENENMEYER, J.H., DREW, L.J. (2011): Statistic for earth and environmental scientists. Wiley, Singapore.

SENN, D.B., HEMOND, H.F. (2002): Nitrate controls on iron and arsenic in an urban lake, Science, 296, 2373-2376.

SENGUPTA, S., SARKAR, A. (2006): Stable isotope evidence of dual (Arabian Sea and Bay of Bengal) vapour sources in monsoonal precipitation over north India. Earth Planet. Sci. Lets., 250, 511-521.

SENGUPTA, S., MC ARTHUR, J.M., SARKAR, A., LENG, M.J., RAVENSCROFT, P., HOWARTH, R.J., BANERJEE, D.M. (2008): Do Ponds Cause Arsenic-Pollution of Groundwater in the Bengal Basin? An Answer from West Bengal. Environ. Sci. Technol., 42, 5156-5164.

SHANLEY, K.W., MC CABE, P.J. (1994): Perspectives on the sequence stratigraphy of continental strata. Am. Assoc. Petrol. Geol. Bull., 78, 544-568. In: MIAL, A.D., Ed. (1996): The Geology of Fluvial Deposits. Springer, Berlin, Heidelberg, New York.

SHARP, Z. (2006): Principles of stable isotope geochemistry. Prentice Hall, New Jersey.

SMEDLEY, P.L., KINNIBURGH, D.G. (2002): A review of the source, behaviour and distribution of arsenic in natural waters. Appl. Geochem., 17, 517–568.

SMITH, A.H., LINGAS, E.O., RAHMAN, M. (2000): Contamination of drinking-water by Arsenic in Bangladesh: a public health emergency. Bulletin of the World Health Organization, 78, 1093-1103.

STEINFINK, H., SANS, F.J. (1959): Refinement of the crystal structure of dolomite. Am. Min., 44, 679-682.

REFERENCES

STEINFINK, H. (1962): Crystal structure of a trioctahedral mica: Phlogopite. Am. Min., 47, 886-889.

STOLLENWERK, K.G., BREIT, G.N., WELCH, A.H., YOUNT, J.C., WHITNEY, J.W., FOSTER, A.L., UDDIN, M.N., MAJUMDER, R.K., AHMED, N. (2007): Arsenic attenuation by oxidized aquifer sediments in Bangladesh. Sci. Tot. Environ., 379, 133-150.

STOYAN, D., STOYAN, H., JANSEN, U. (1997): Umweltstatistik, Statistische Verarbeitung und Analyse von Umweltdaten. 56, B.G. Teubner Verlagsgesellschaft, Stuttgart, Leipzig.

STRIBLING, J.M., CORNWELL, J.C. (1997): Identification of Important Primary Producers in a Chesapeake Bay Tidal Creek System Using Stable Isotopes of Carbon and Sulfur. Estuaries, 20 (1), 77-85.

STUMM, W., MORGAN, J.J. (1996): Aquatic chemistry: Chemical equilibria and rates in natural waters, 3rd edition. John Wiley & Sons, N.Y.

STÜBEN, D., BERNER, Z., CHANDRASEKHRARAM, D, KARMAKAR, J. (2003): Arsenic enrichment in groundwater of West Bengal, India: Geochemical evidence for mobilization of As under reducing conditions. Appl. Geochem., 18, 1417-1434.

STUTE, M., ZHENG, Y., SCHLOSSER, P., HORNEMAN, A., DHAR, R.K., DATTA, S., HOQUE, M.A., SEDDIQUE, A.A., SHAMSUDDUHA, M., AHMED, K.M., VAN GEEN, A. (2007): Hydrological control on As concentration in Bangladesh groundwater. Water Resour. Res., 43, W09417, DOI: 10.1029/2005SWR004499.

SUTTON, N.B., VAN DER KRAAN, G.M., VAN LOOSDRECHT, M.C., MUYZER, G., BRUINING, J., SCHOTTING, R.J. (2009): Characterization of geochemical constituents and bacterial populations associated with As mobilization in deep and shallow tube wells in Bangladesh. Water Res. 2009, 43, 1720-1730.

TUFANO, K.J., FENDORF, S. (2008): Confounding impacts of iron reduction on arsenic retention. Environ. Sci. Technol., 42, 4777-4783.

UNDP (1982): Groundwater survey- The hydrogeologic conditions of Bangladesh. UNDP Technical Report, DP/UN/BGD-74-009/1, New York.

UNESCO (2009): The United Nations World Water Development Report 3: Water in a changing world. UNESCO publishing, Paris, London. Available from: http://unesdoc.unesco.org/ images/0018/001819/181993e.pdf.

VAIDYANATHAN, G. (2011): Arsenic sinks to new depths. Nature news, DOI:10.1038/news.2011.20.

VAN GEEN, A., ZHENG, Y., STUTE, M., AHMED, K.M. (2003): Comment on "Arsenic Mobility and Groundwater Extraction in Bangladesh" (II). Science: 584, DOI:10.1126/science.1081057.

VAN GEEN, A., ROSE, J., THORAL, S., GARNIER, J.M., ZHENG, Y., BOTTERO, J.Y. (2004): Decoupling of As and Fe release to Bangladesh groundwater under reducing conditions. Part II: Evidence from sediment incubations. Geochim. Cosmochim. Acta, 68, 3475–3486.

VAN GEEN, A., ZHANG, Y., CHENG, Z., AZIZ, Z., HORNEMAN, A., DHAR, R.K., MAILLOUX, B., STUTE, M., WEINMAND, B., GOODBRED, S., SEDDIQUE, A.A. (2006): A transect of groundwater and sediment properties in Araihazar, Bangladesh: Further evidence of decoupling between As and Fe mobilization. Chem. Geol., 228, 85-96.

VAN GEEN, A., CHENG, Z., JIA, Q., SEDDIQUE, A.A., RAHMAN, M.M., AHMED, K.M. (2007): Monitoring 51 community wells in Araihazar, Bangladesh, for up to 5 years: Implications for arsenic mitigation. J. Environ. Sci. Health Pt. A, 42, 1729–1740.

VAN GEEN, A. (2008): Arsenic meets dense populations. Nat. Geosci., 1, 494-496.

VAN GEEN, A., RADLOFF, K.A., AZIZ, Z., CHENG, Z., HUQ, M.R., AHMED, K.M., WEINMAN, B., GOODBRED, S., JUNG, H.B., ZHENG, Y., BERG, M., TRANG, P.H.K., CHARLET, L., MÉTRAL, J., TISSERAND, D., GUILLOT, S., CHAKRABORTY, S., GAJUREL, A.P., UPRETI, B.N. (2008): Comparison of arsenic concentrations in simultaneously-collected ground water and aquifer particles from Bangladesh, India, Vietnam, and Nepal. Appl. Geochem., 23, 3244-3251.

VAN GEEN (2011): International Drilling to Recover Aquifer Sands (IDRAs) and Arsenic Contaminated Groundwater in Asia. Sci. Drilling, 12, 49-52.

VAN HERREWEGHE, S., SWENNEN, R., VANDECASTEELE, C., CAPPUYNS, V. (2003): Solid phase speciation of arsenic by sequential extraction in standard reference materials and industrially contaminated soil samples. Environ. Pollut., 122, 323-342.

REFERENCES

WAGNER, F., BERNER, Z., STÜBEN, D. (2005): Arsenic in groundwater of the Bengal Delta Plain: Geochemical evidences for small scale redox zonation in the aquifer. In: BUNDSCHUH, J., BHATTARACHARYA, P., CHANDRASEKHARAM, D., Eds. (2005): Natural Arsenic in Groundwater: Occurrences, remediation and management. 3-15, Taylor & Francis Group, London, UK.

WAGNER, F. (2005): Prozessverständnis einer Naturkatastrophe: eine geo- und hydrochemische Untersuchung der regionalen Arsen-Anreicherung im Grundwasser West-Bengalens (Indien). Karlsruher Mineralogische und Geochemische Hefte. Schriftenreihe des Instituts für Mineralogie und Geochemie, Universität Kalrsruhe (TH), Band 29, Universitätsverlag Karlsruhe.

WARD JR., J.H. (1963): Hierarchical Grouping to Optimize an Objective Function. J. Am. Stat. Ass., 58 (301), 236-244.

WASSERMANN, G.A., LIU, X., PARVEZ, F., AHSAN, H., FACTOR-LITVAK, P., VAN GEEN, A., CHENG, Z., SLAVOKOVICH, V., HUSSAIN, I., MOMOTAJ, H., GRAZIANO, J.H. (2004): Water arsenic exposure and children's intellectual function in Araihazar, Bangladesh. Environ. Health Perspect., 112, 1329-1333.

WEHRLI, B., STUMM, B. (1988): Vanadyl in natural waters: Adsorption and hydrolysis promote oxygenation. Geochim. Cosmochim. Acta, 53, 69-77.

WELCH, A.H., WESTJOHN, D.B., HELSEL, D.R., WANTY, R.B. (2000): Arsenic in groundwater of the United States: Occurrences and geochemistry. Ground Water 38, 589–604. In: DIXIT, S., HERING, J.G. (2006): Sorption of Fe(II) and As(III) on goethite in single- and dual-sorbate systems. Chem. Geol., 228, 6-15.

WENZEL, W.W., KIRCHBAUMER, N., PROHASKA, T., STINGEDER, G., LOMBIC, E., ADRIANO, D.C. (2001): Arsenic fractionation in soils using an improved sequential extraction procedure. Anal. Chim. Acta, 436, 309-323.

WHITEHEAD, T.H., BRADBURY, W.C. (1950): Qualitative Scheme of Analysis for the Common Sugars. Anal. Chem., 22, 651–653.

WHO (2003): Arsenic in Drinking-water - Background document for development of WHO Guidelines for Drinking-water Quality. WHO/SDE/WSH/03.04/75. Available from: http://www.who.int/ water_sanitation_health/dwq/chemicals/arsenic.pdf.

WHO (2011): Guidelines for Drinking-water Quality. 4th edition. World Health Organization, Geneva. Available from: http://whqlibdoc.who.int/publications/ 2011/9789241548151_eng.pdf.

WINKEL, L.H.E., BERG, M., AMINI, M., HUG, S.J., JOHNSON, A. (2008): Predicting groundwater arsenic contamination in Southeast Asia from surface parameters. Nat. Geosci., 1, 536-542.

WINKEL, L.H.E., TRANG, P.T.K., LAN, V.M., STENGEL, C., AMINI, M., HA, N.T., VIET, P.H., BERG, M. (2011): Arsenic pollution of groundwater in Vietnam exacerbated by deep aquifer exploitation for more than a century. PNAS, 108, 1246-1251.

WOLF, M., KAPPLER, A., JIIANG, J., MECKENSTOCK, R.U. (2009): Effects of humic substances and quinones at low concentrations on ferrihydrite reduction by *geobacter metallireducens*. Environ. Sci. Technol., 43, 5679-5685.

WORLD BANK (2005): Report No.31303. Towards a more effective operational response – Arsenic contamination of groundwater in South and East Asian countries. Volume II Technical Report. Available from: http://www-wds.worldbank.org/servlet/WDSContentServer/WDSP/IB/2005/04/06/ 000090341_20050406130256/Rendered/PDF/313030v1.pdf.

WRIGHT, V.P., MARRIOTT, S.B. (1993): The sequence stratigraphy of fluvial depositional systems: the role of floodplain sediment storage. Sed. Geol., 86, 203-210.

YANINA, S.V., ROSSO, K.M. (2008): Linked Reactivity at Mineral-Water Interfaces Through Bulk Crystal Conduction. Science, 320, 218-222.

YU, W.H., HARVEY, C.M., HARVEY, C.F. (2003): Arsenic in groundwater in Bangladesh: A geostatistical and epidemiological framework for evaluating health effects and potential remedies. Water Resour. Res., 39 (6), 1-17.

ZACHARA, J.M., FREDRICKSON, J.K., SMITH, S.C., GASSMAN, P.L. (2001): Solubilization of Fe(III) oxide–bound trace metals by a dissimilatory Fe(III)-reducing bacterium. Geochim. Cosmochim. Acta, 65, 75–93.

ZOBRIST, J., DOWDLE, P.R., DAVIS, J.A., OREMLAND, R.S. (2000): Mobilization of Arsenite by Dissimilatory Reduction of Adsorbed Arsenate. Environ. Sci. Technol., 34, 4747-4753.

APPENDIX

A.1   MATERIALS AND METHODS

Table A 1.1, part I: Average EDX measuring results of reference materials (<u>recommended value</u>, proposed value, *information value*; GLADNEY & ROELANDTS 1990, GOVINDARAJU 1994). Sample contents were divided by the correction value, which represents the average recovery rate (rec.) of the reference materials. Outliers (*) were excluded from the correction factor calculation. Table continued on next page.

Std.	Value	K2O (wt.%)	CaO (wt.%)	TiO_2 (wt.%)	MnO (wt.%)	Fe_2O_3 (wt.%)	Ni (mg/kg)	Cu (mg/kg)	Zn (mg/kg)	As (mg/kg)
Soil V	Avg.	2.07	3.44	0.79	0.12	6.40	21.6	83.3	378 11	98.0
n: 10	±	0.11	0.05	0.01	0.00	0.16	0.55	2.2		2.8
	Ref.	<u>2.24</u>	3.10	<u>0.78</u>	<u>0.11</u>	6.36	13.0	<u>77.0</u>	<u>370</u>	94.0
	Rec.	0.92	1.11	1.02	1.06	1.01	1.66*	1.08	1.02	1.04
Soil VII	Avg.	1.67	30.2	0.65	0.11	4.86	30.5	17.7	124 1	17.7
n: 10	±	0.01	0.09	0.00	0.00	0.02	0.98	0.6		0.4
	Ref.	-	-	-	<u>0.08</u>	-	-	<u>11.0</u>	<u>104</u>	13.4
	Rec.				1.33			1.61*	1.19	1.32
GXR-2	Avg.	1.49	1.40	0.61	0.15	3.13	29.3	88.5	595	24.9
n: 11	±	0.01	0.01	0.00	0.00	0.01	0.4	0.9	2	0.8
	Ref.	1.65	1.30	0.50	0.13	2.66	21.0	76.0	530	25.0
	Rec.	0.90	1.08	1.22	1.14	1.18	1.40*	1.16	1.12	1.00
GXR-5	Avg.	0.79	1.23	0.49	0.04	5.26	64.5	369	48.2	12.2
n: 10	±	0.01	0.01	0.00	0.00	0.01	0.6	1	0.8	0.3
	Ref.	1.06	1.17	0.37	0.04	4.84	75.0	354	49.0	11.2
	Rec.	0.75*	1.05	1.33	1.04	1.09	0.86	1.04	0.98	1.09
Sco-1	Avg.	2.59	3.07	0.58	0.05	5.32	33.1	29.7	102	13.0
	±	0.02	0.01	0.00	0.00	0.01	0.9	0.6	1	0.3
n: 7	Ref. value	<u>2.77</u>	<u>2.62</u>	0.63	0.05	5.14	27.0	28.7	<u>103</u>	12.4
	Rec.	0.93	1.17	0.92	0.97	1.03	1.23	1.03	0.99	1.04
SDO-1	Avg.	3.09	1.16	0.65	0.04	8.73	75.7	52.4	54.3	58.7
n: 7	±	0.16	0.03	0.02	0.00	0.27	2.6	1.5	1.6	2.4
	Ref.	<u>3.35</u>	<u>1.05</u>	<u>0.71</u>	<u>0.04</u>	<u>9.34</u>	<u>99.5</u>	60.2	<u>64.1</u>	68.5
	Rec.	0.92	1.11	0.91	0.96	0.93	0.76	0.87	0.85	0.86
Correction value		0.88	1.10	1.08	1.08	1.05	0.95	1.04	1.03	1.01

Table A 1.1, part II.

Std.	Value	Rb	Sr	Y	Zr	Nb	Mo	Ba	La	Ce	Pb
						all (mg/kg)					
Soil V	Avg.	124	322	22.5	210	9.81	2.31	593	31.2	56.9	134
n: 10	±	5	11	1.3	11	0.56	0.39	10	1.3	1.5	5
	Ref.	<u>140</u>	330	21.0	220	9.00	2.00	<u>560</u>	<u>28.0</u>	<u>60.0</u>	<u>130</u>
	Rec.	0.89	0.98	1.07	0.95	1.09	1.15	1.06	1.11	0.95	1.03
Soil VII	Avg.	59.6	27.4	126	212	12.4	1.62	163	32.3	57.2	67.3
n: 10	±	0.9	1.2	1	2	0.7	0.50	2	0.9	1.4	1.0
	Ref.	<u>51.0</u>	21.0	<u>108</u>	<u>185</u>	-	-	-	<u>28.0</u>	<u>61.0</u>	<u>60.0</u>
	Rec.	1.17	1.31	1.17	1.14				1.15	0.94	1.12
GXR-2	Avg.	84.1	164	21.0	282	13.5	1.79	1990	24.7	47.5	716
n: 11	±	1.3	1	0.5	2	0.6	1.16	20	1.3	1.5	11
	Ref.	78.0	160	17.0	269	11.0	*2.10*	2240	25.6	51.4	690
	Rec.	1.08	1.02	1.23	1.05	1.23	0.85	0.89	0.97	0.92	1.04
GXR-5	Avg.	37.2	115	16.4	171	7.25	37.0	2190	18.3	34.0	15.0
n: 10	±	0.7	1	0.2	1	0.32	0.8	20	1.2	1.4	1.3
	Ref.	41.0	110	16.0	140	6.70	31.0	2000	18.9	39.0	21.0
	Rec.	0.91	1.05	1.02	1.22	1.08	1.19	1.09	0.97	0.87	0.72*
Sco-1	Avg.	115	168	25.1	173	12.7	1.27	588	31.5	55.1	28.1
n: 7	±	1	2	0.5	2	0.3	0.46	8	0.9	0.8	1.3
	Ref.	<u>112</u>	<u>174</u>	<u>26.0</u>	<u>160</u>	11.0	<u>1.37</u>	<u>570</u>	<u>29.5</u>	62.0	<u>31.0</u>
	Rec.	1.03	0.96	0.96	1.08	1.15	0.93	1.03	1.07	0.89	0.91
SDO-1	Avg.	114	68.3	38.1	144	12.7	161	381	38.6	68.2	22.6
n: 7	±	8	3.8	2.3	11	1.3	2	11	1.4	2.7	2.6
	Ref.	126	<u>75.1</u>	<u>40.6</u>	<u>165</u>	<u>11.4</u>	<u>134</u>	<u>397</u>	<u>38.5</u>	<u>79.3</u>	<u>27.9</u>
	Rec.	0.90	0.91	0.94	0.87	1.11	1.20	0.96	1.00	0.86	0.81
Correction value		1.00	1.01	1.09	1.05	1.13	1.07	1.01	1.05	0.91	0.98

Table A 1.2, part I: Comparison of an in-house quartz reference material with literature values of principal peaks with relative intensities (RI) >5% between 3 and 63° 2θ and comparison of duplicate measurements for sediment samples (RI >1.50 %) to prove the reliability of XRD analysis. Continued on next page.

Quartz				Sample	High As site, 4.60 m bls			
LEVIEN et al. (1980)		In-house reference			1		2	
d (Å)	RI (%)	d (Å)	RI (%)	Peak	d (Å)	RI (%)	d (Å)	RI (%)
3.3446	100	3.3457	100	1	3.3414	100	3.3415	100
4.2574	28.5	4.2614	11.6	2	1.8175	20.4	1.8175	20.5
1.8184	18.0	1.8189	15.6	3	1.5413	16.0	1.5413	16.0
1.5423	13.0	1.5420	14.1	4	3.1844	15.2	3.1852	15.0
2.4580	10.2	2.4584	7.89	5	4.2543	13.8	4.2544	14.1
2.2818	8.84	2.2827	8.81	6	2.4560	9.83	2.4562	9.78
2.1287	7.51	2.1288	7.98	7	2.1273	9.64	2.1270	9.70
				8	2.2811	8.25	2.2806	8.53
				9	3.0267	7.68	3.0251	8.30
				10	1.6713	6.28	1.6713	6.74
				11	1.9925	6.19	1.9923	6.28
				12	2.8842	6.13	2.8841	6.08
				13	1.9797	5.20	1.9797	5.31
				14	2.2360	5.02	2.2362	4.84
				15	9.9878	4.51	9.9815	4.55
				16	3.2419	4.51	3.2411	4.60
				17	3.5339	2.67	3.5337	2.75
				18	1.6593	2.54	1.6591	2.86
				19	4.9791	2.52	4.9801	2.57
				20	2.5641	2.50	2.5641	2.65
				21	3.4835	2.40	3.4951	2.35
				22	1.7838	2.33	1.7838	2.02
				23	2.9905	2.22	2.9874	2.16
				24	2.4906	2.04	2.4903	1.89
				25	1.8040	2.02	1.8027	2.18
				26	7.0800	1.79	7.0742	1.78
				27	3.7698	1.78	3.7689	1.87
				28	2.8591	1.70		
				29	1.9097	1.68	1.9092	1.63
				30	4.0323	1.65	4.0313	1.67

Table A 1.2, part II.

Sample	High As site, 10.5 m bls				Low As site, 11.4 m bls			
	1		2		1		2	
Peak	d (Å)	RI (%)	d (Å)	RI (%)	d (Å)	RI (%)	d (Å)	RI (%)
1	3.3397	100	3.3410	100	3.3395	100	3.3407	100
2	1.8169	20.3	1.8174	19.5	1.8170	17.7	1.8170	16.8
3	1.5407	15.9	1.5411	17.1	1.5409	15.4	1.5411	13.5
4	3.1861	14.5	3.1823	16.8	4.2511	12.9	4.2514	14.7
5	4.2513	13.0	4.2522	15.1	3.1854	11.7	3.1874	14.2
6	2.4545	10.3	2.4561	8.82	2.4550	9.31	2.4556	6.53
7	2.8842	10.9	2.8852	3.89	2.2799	8.90	2.2796	7.67
8	2.2793	8.43	2.2807	7.45	2.1264	8.45	2.1262	9.78
9	3.2418	7.73	3.2397	9.40	1.9786	6.74	1.9789	5.03
10	2.1259	7.64	2.1263	13.2	1.6711	5.59	1.6710	6.65
11	1.6706	6.74	1.6712	7.25	2.2352	4.82		
12	1.9785	5.85	1.9797	5.85	3.2402	4.68	3.2308	7.08
13	1.9917	4.66	1.9928	5.65	3.0274	4.09	3.0258	5.18
14	3.0239	4.12	3.0240	9.36	2.8859	4.06	2.8818	5.05
15	2.2351	3.98	2.2361	3.53	1.9911	3.17	1.9922	5.31
16	9.9873	3.41	9.9887	4.89	9.9694	3.00	9.9968	3.70
17	1.6582	2.18	1.6590	2.78	1.6586	2.68	1.6587	2.41
18	3.7650	2.05	3.7679	1.68	3.7660	2.19		
19	3.4799	1.95	3.4869	1.85	1.8004	2.09	1.7986	2.02
20	4.9727	1.71	4.9804	2.20	3.4794	1.81	3.4803	1.98
21	3.5288	1.69	3.5325	1.84	1.7841	1.76		
22	1.8013	1.58	1.8007	1.61	2.9886	1.62		

Table A 1.3: Magnetic separation settings and characteristic minerals in respective fractions (according to instruction manual).

Fraction	Inclination - front (°)	Inclination - lateral (°)	Magnetic field current (A)	Characteristic minerals
0	Bar magnet	-	-	Magnetite
1	30	20	0.4	Garnet, ilmenite, olivine, chromite, chloritoid
2	30	20	0.8	Hornblende, hypersthene, augite, actinolite, staurolite, epidote, biotite, chlorite, tourmaline (dark)
3	30	20	1.2	Tremolite, enstatite, spinel, staurolite (light), muscovite, zoisite, clinozoisite, tourmaline (light)
4	30	5	1.2	Apatite, andalusite, monazite, xenotine
5	Residuum	-	-	Zircon, rutile, anatase, pyrite, corundum, topaz, fluorite, silimanite, anhydrite, beryl

Table A 1.4: Sequential extraction procedure scheme proposed by EICHE et al. (2008). Although being semi-quantitative, this procedure provides a reliable estimation of the sedimentary As distribution among different potential host phases. After completion of each fraction, a wash step with Milli-Q water was done (volume similar to extractant; fraction VI: hot wash) and included in the respective extraction solutions. Solutions were centrifuged (15 min at 4500 rpm), decanted and stored in a refrigerator at 4°C until analysis by HR-ICP-MS.

Step	Target phase	Extractant	Conditions
I	Weakly (electro-statically) bound As	0.05 M $(NH_4)_2SO_4$ (p.a., Merck)	Volume: 25 mL Leaching duration: 4 h Temperature: 25°C Repetition: 1
II	Strongly adsorbed (ligand exchange) As	0.5 M NaH_2PO_4 (p.a., Merck)	Volume: 40 mL Leaching duration: 16 & 24 h Temperature: 25°C Repetition: 1 (each)
III	Arsenic incorporated in acid volatile sulphides (AVS), carbonates, Mn-oxides, very amorphous Fe-(oxyhydr)oxides	1 M HCl (p.a., Merck)	Volume: 40 mL Leaching duration: 1 h Temperature: 25°C Repetition: 1
IV	Arsenic incorporated in amorphous Fe-(oxyhydr)oxides	AOD: 0.2 M NH_4-oxalate (p.a., Merck) + 0.2 M oxalic acid (p.a., ROTH) → Forms complexes with Fe that are sensitive to light	Volume: 40 mL Leaching duration: 2 h Temperature: 25°C Repetition: 1 Solutions and extractions kept in dark
V	Arsenic incorporated in crystalline Fe-(oxyhydr)oxides	DCB: 0.5 M Na-citrate (p.a., Merck) + 1 M $NaHCO_3$ (p.a., Merck) after heating to 85°C: + 0.5 g $Na_2S_2O_4 \times H_2O$ (p.a., Merck)	Volume: 35 mL (Na-citrate) + 2.5 ml ($NaHCO_3$) + 0.5 g ($Na_2S_2O_4 \times H_2O$) Leaching duration: 15 min Temperature: 85°C Repetition: 1
VI	Arsenic associated with silicates	10 M HF (suprapure, Merck) after 16 h: + 5 g boric acid (p.a., Merck)	Volume: 40 mL Leaching duration: 1 & 24 h Temperature: 25°C Repetition: 1 (each)
VII	Arsenic-bearing sulphides, refractory minerals, OM	16 M HNO_3 (sub-boiled, Merck) + H_2O_2 (30% p.a., ROTH)	According to EPA 3050B

Table A 1.5: Reference materials used for sedimentary organic matter characterisations (*reference values cited from IAEA 1993, **SCHOELL et al. 1983, otherwise cited from respective manufacturer certificates).

Instrument	Parameter	Ref. material	Ref. value	Measured average	n
CSA	TOC / TIC / TS	ELTRA Steel standard 92000-27	C: 0.154 ± 0.003 (wt.%) S: 0.026 ± 0.0001 (wt.%)	C: 0.155 ± 0.002 wt.% S: 0.028 ± 0.002 wt.%	30
		ELTRA Steel standard 92000-26	C: 3.21 ± 0.02 (wt.%) S: 0.127 ± 0.004 (wt.%)	C: 3.296 ± 0.015 wt.% S: 0.124 ± 0.002 wt.%	3
EA	TN	Heka Tech Soil 5	0.021 (wt.%)	0.025 ± 0.003 wt.%	15
		NCS DC 73326	0.037 ± 0.001 (wt.%)	0.038 ± 0.003 wt.%	24
EA-IRMS	δ^{15}N	IAEA-N1*	0.43 ± 0.07 (‰ air)	0.40 ± 0.13 (‰ air)	69
		IAEA-N2*	20.32 ± 0.09 (‰ air)	20.31 ± 0.14 (‰ air)	26
		25-USGS*	-30.25 ± 0.38 (‰ air)	-30.38 ± 0.23 (‰ air)	29
EA-IRMS	δ^{13}C	NBS-21**	-28.16 ± 0.01 (‰ VPDB)	-28.23 ± 0.13 (‰ VPDB)	38
		USGS 24*	-15.994 ± 0.105 (‰ VPDB)	-15.96 ± 0.05 (‰ VPDB)	12
		NBS-18*	-5.029 ± 0.049 (‰ VPDB)	-5.09 ± 0.08 (‰ VPDB)	11

Table A 1.6: Determined elements by HR-ICP-MS measured at given masses and average ld (3*σ, in µg L^{-1}) from all monitoring and in-situ experiment analyses (n: 13).

^{23}Na	^{26}Mg	^{39}K	^{42}Ca	^{43}Ca	^{7}Li	^{27}Al
22.6 ± 9.5	6.62 ± 3.45	22.7 ± 10.6	27.1 ± 14.1	25.3 ± 14.1	0.15 ± 0.06	0.43 ± 0.20

^{31}P	^{47}Ti	^{51}V	^{52}Cr	^{55}Mn	^{56}Fe	^{59}Co
0.81 ± 0.38	0.07 ± 0.07	0.01 ± 0.01	0.02 ± 0.02	0.22 ± 0.09	1.55 ± 0.72	0.03 ± 0.01

^{60}Ni	^{65}Cu	^{66}Zn	^{75}As	^{75}As(III)	^{85}Rb	^{88}Sr
0.17 ± 0.19	0.23 ± 0.24	0.72 ± 0.40	0.25 ± 0.08	0.39 ± 0.22	0.03 ± 0.02	0.12 ± 0.06

^{98}Mo	^{111}Cd	^{133}Cs	^{137}Ba	^{205}Tl	^{207}Pb	^{238}U
0.13 ± 0.08	0.02 ± 0.03	0.02 ± 0.01	0.08 ± 0.05	0.03 ± 0.01	0.04 ± 0.02	0.00 ± 0.01

Table A 1.7: Certified reference solutions used for HR-ICP-MS measurements of all monitoring and in-situ experiment samples (A: Trace Metals In Drinking Water, HPS; B: Trace Metals ICP QCP 050-1, Promochem GmbH).

Element	As* (µg/L)	Na (mg/L)	Mg (mg/L)	K (mg/L)	Ca (mg/L)	Li (µg/L)
A	11.0	0.50	1.60	0.50	6.20	3.00
Analysis	11.2	0.46	1.59	0.50	6.18	3.11
(n: 29)	± 0.3	± 0.03	± 0.07	± 0.03	± 0.20	± 0.23
B		0.16	0.02	0.78	0.02	12.6
Analysis		0.17	0.02	0.73	0.03	12.7
(n: 7)		± 0.02	± 0.00	± 0.05	± 0.01	± 1.2

Element	Al (µg/L)	Ti (µg/L)	V (µg/L)	Cr (µg/L)	Mn (µg/L)	Fe (µg/L)
A	25.0	-	7.00	4.00	8.00	18.0
Analysis	24.8	-	7.10	4.06	8.14	19.2
(n: 29)	± 1.2		± 0.28	± 0.14	± 0.31	± 1.6
B	116	14.7	18.9	8.40	8.40	6.30
Analysis	113	14.8	19.0	8.41	8.40	6.88
(n: 7)	± 3	± 0.5	± 0.7	± 0.1	± 0.19	± 0.61

Element	Co (µg/L)	Ni (µg/L)	Cu (µg/L)	Zn (µg/L)	Sr (µg/L)	Mo (µg/L)
A	5.00	12.0	4.00	15.0	60.0	22.0
Analysis	4.96	12.1	4.12	15.3	60.2	21.9
(n: 29)	± 0.19	± 0.5	± 0.30	± 0.5	± 1.7	± 0.8
B	12.6	14.7	14.7	6.30	29.4	18.9
Analysis	12.2	14.6	14.5	6.51	30.3	19.2
(n: 7)	± 0.4	± 0.3	± 0.4	± 0.32	± 0.9	± 0.5

Element	Cd (µg/L)	Ba (µg/L)	Tl (µg/L)	Pb (µg/L)
A	2.00	100	2.00	4.00
Analysis	2.00	100	2.00	3.90
(n: 29)	± 0.07	± 3	± 0.04	± 0.15
B	21.0	137	27.3	14.7
Analysis	21.4	139	27.7	15.0
(n: 7)	± 0.8	± 4	± 0.7	± 0.7

*As(III) determination: 11.5 ± 0.5, n: 11 (determined from reference solution A)

Table A 1.8: Quantification limits (confidence interval: 0.99) for anions determined in all monitoring and in-situ experiment samples by IC (column 1: DIONEX AS 4 SC; column 2: DIONEX AS14). In brackets: ranges of five point calibration (SPEC PURE™ Multi Ion IC Standard Solution, Alfa-Aesar).

Column	n	Cl^- (mg/L)	NO_3^- (mg/L)	SO_4^{2-} (mg/L)
Column 1 (used until February 2010)	28	0.26 (0.4 to 20.0)	0.88 (0.8 to 40.0)	0.85 (0.8 to 40.0)
Column 2 (used after February 2010)	9	0.36 (1.0 to 40.0)	0.28 (2.0 to 80.0)	0.37 (2.0 to 80.0)
# Reference solutions		4	8	8

Table A 1.9: Precision for isotopic reference materials, measured by IRMS against VSMOW (Vienna Standard Mean Ocean Water) during in-situ experiment sample analyses (reference values by IAEA 1993).

Parameter	Reference material	n	Reference value	Analysis
$\delta^{18}O$	GISP*	3	-24.78	-24.72 ± 0.08
	SLAP**	3	-55.50	-55.53± 0.03
	VSMOW***	3	0	-0.04 ± 0.04
δ^2H	GISP	7	-189.73	-189.9 ± 1.2
	Slap	7	-428.00	-427.80 ± 0.54
	VSMOW	7	0	0.23 ± 0.68

*GISP: Greenland Ice Sheet Precipitation

**SLAP: Standard Light Antarctic Precipitation

***VSMOW: Vienna Standard Mean Ocean Water

A.2   THE LOW ARSENIC STUDY SITE

Table A 2.1: Summary of sediment analysis, comprising all samples down to a depth of 45.5 m bls (n: 70).

Value range	K_2O (wt.%)	CaO (wt.%)	TiO_2 (wt.%)	MnO (wt.%)	Fe_2O_3 (wt.%)	As (mg/kg)	Ni (mg/kg)	Cu (mg/kg)	Zn (mg/kg)
Minimum	1.37	0.97	0.20	0.02	1.40	1.8	21.2	8.73	19.0
25 % Q.	1.87	2.60	0.26	0.03	2.09	2.8	23.3	9.76	27.2
Median	2.31	2.98	0.34	0.04	2.37	2.9	24.2	10.4	30.3
75 % Q.	2.76	3.80	0.48	0.05	3.24	3.6	29.2	13.6	45.2
Maximum	4.00	5.33	0.83	0.09	8.06	12.6	59.1	59.9	109
Average	2.35	3.16	0.39	0.04	2.88	3.7	27.1	14.0	37.6
r_{As-}	0.64	-0.11	0.75	0.67	0.86		0.89	0.92	0.87
r_{Fe-}	0.78	0.07	0.93	0.78		0.86	0.97	0.94	0.99

Value range	Ba (mg/kg)	Sr (mg/kg)	Ce (mg/kg)	Y (mg/kg)	La (mg/kg)	Fine fraction* (wt.%)	TS (wt.%)	TOC (wt.%)	TIC (wt.%)
Minimum	243	75.1	38.1	11.7	18.0	0.19	<0.003	0.02	0.02
25 % Q.	310	85.9	49.6	15.3	25.4	2.82	0.005	0.02	0.31
Median	325	92.5	58.0	17.9	29.5	10.6	0.006	0.03	0.38
75 % Q.	367	99.9	78.4	24.1	37.7	18.9	0.009	0.07	0.52
Maximum	581	109	98.6	34.6	47.3	94.2	0.064	1.67	1.33
Average	345	92.2	63.7	19.8	31.1	19.2	0.008	0.09	0.44
r_{As-}	0.83	0.07	0.29	0.50	0.27	0.76	0.53	0.75	0.24
r_{Fe-}	0.93	0.22	0.48	0.66	0.48	0.88	0.40	0.51	0.36

*Fraction <0.063 mm, comprising silt and clay

Table A 2.2: Distribution of As, Fe and Mn in SEP fractions I-VII of representative samples (fractions described in Table A 1.4). GXR-2 reference values (mg/kg): As: 25.0; Fe: 19,000; Mn: 1,007.

Depth (m bls)	As (mg/kg)							
	I	II	III	IV	V	VI	VII	Σ
2.55	0.16	4.94	0.55	0.60	1.07	1.67	0.06	9.05
3.20	0.06	1.73	0.38	0.20	0.03	0.00	0.23	2.62
13.0	0.05	1.67	0.19	0.19	0.04	0.00	0.62	2.76
20.2 A	0.05	0.68	0.23	0.16	0.36	0.00	0.00	1.49
20.2 B	0.02	1.09	0.33	0.06	0.14	0.00	0.00	1.63
25.3	0.01	0.50	0.29	0.11	0.27	0.00	0.00	1.18
30.5	0.01	2.53	0.30	0.13	0.00	0.00	0.00	2.99
38.3	0.00	1.54	0.29	0.06	0.00	0.00	0.00	1.88
45.5	0.01	0.96	0.23	0.04	0.10	0.00	0.00	1.34
GXR-2	1.21	21.4	2.34	2.11	1.38	0.00	0.35	28.8

Depth (m bls)	Fe (mg/kg)							
	I	II	III	IV	V	VI	VII	Σ
2.55	558	647	1914	1409	12469	29885	38.8	46919
3.20	155	3276	6579	1058	3522	17425	119	32134
13.0	17.4	2774	2973	1106	889	9895	432	18086
20.2 A	7.44	782	972	150	561	6051	170	8695
20.2 B	3.41	594	910	197	699	8014	96.4	10513
25.3	7.67	455	899	229	4560	7698	0.00	13849
30.5	4.09	1406	2182	402	1046	6901	45.1	11986
38.3	23.4	1364	1702	250	574	5021	2.05	8936
45.5	5.81	465	1837	247	2079	8108	0.00	12742
GXR-2	169	3175	6362	1698	6043	5043	95.5	22586

Depth (m bls)	Mn (mg/kg)							
	I	II	III	IV	V	VI	VII	Σ
2.55	2.56	135	119	506	89.8	211	1.00	1065
3.20	28.1	167	73.7	174	19.4	182	4.71	649
13.0	17.6	52.3	46.9	1.87	3.98	174	33.3	330
20.2 A	20.3	35.2	23.5	57.2	2.82	110	10.9	260
20.2 B	13.7	31.6	27.5	46.2	3.17	142	6.36	270
25.3	9.25	27.3	32.4	120	34.6	117	7.66	348
30.5	25.1	50.1	37.6	147	9.13	110	3.65	383
38.3	14.9	46.2	29.6	143	4.35	68.3	1.99	309
45.5	12.5	30.1	43.3	105	13.8	121	10.2	336
GXR-2	149	886	72.1	36.6	28.8	76.2	1.55	1250

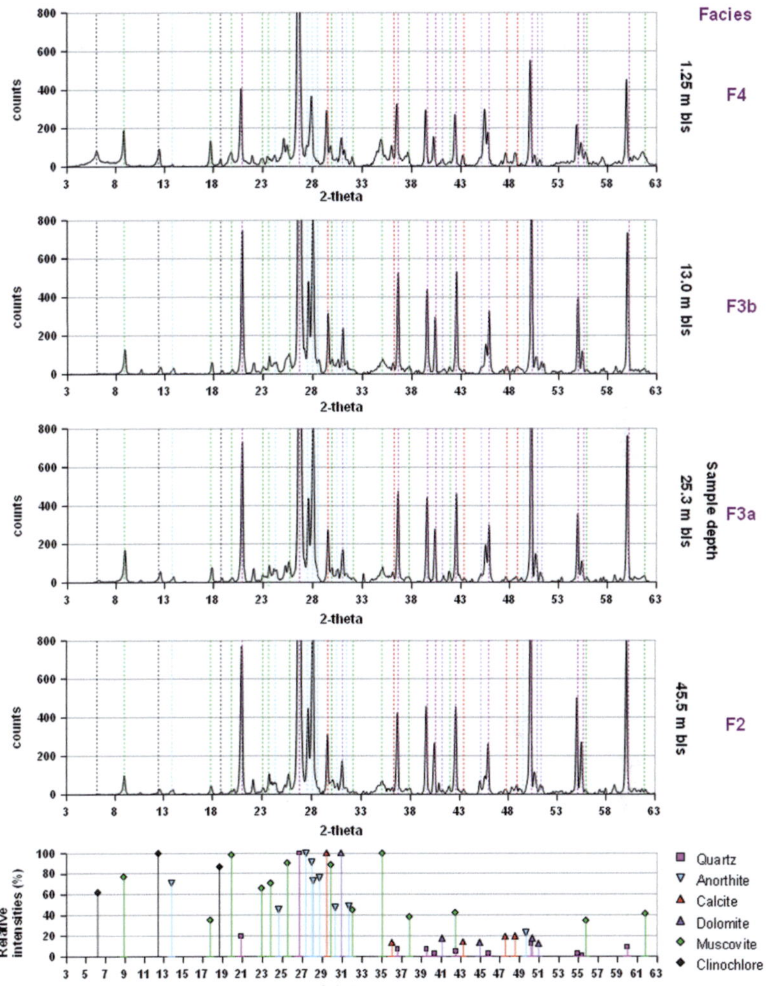

Figure A 2.1: Interpretation of XRD spectra from different depths and litho-facies. Reference spectra of identified mineral peaks include anorthite (ANGEL 1988), calcite (GRAF 1961), clinochlore (PHILIPS et al. 1980), dolomite (STEINFINK & SANS 1959), muscovite (RICHARDSON & RICHARDSON 1982) and quartz (LEVIEN et al. 1980).

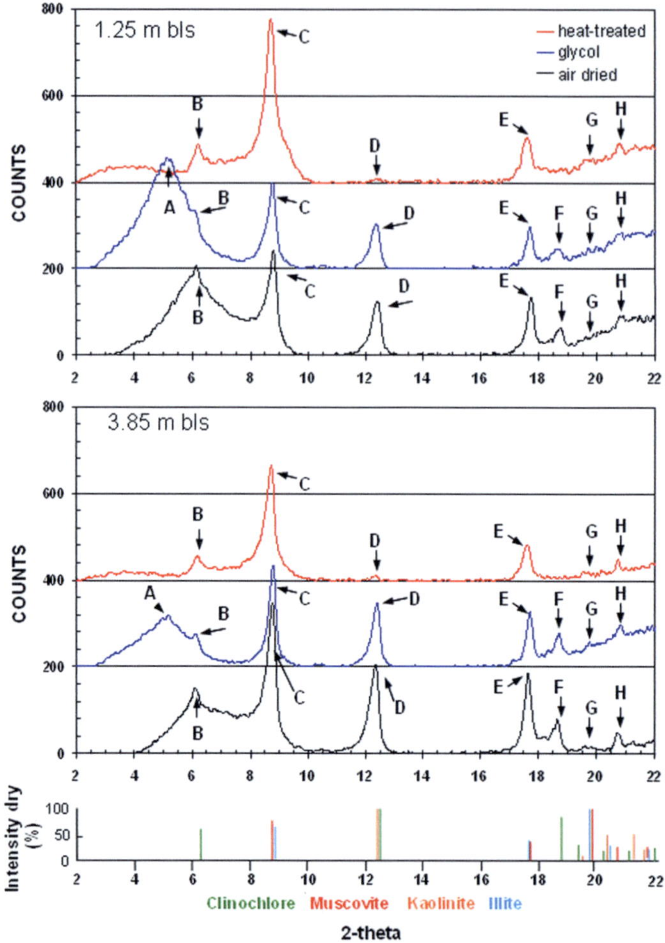

Figure A 2.2, part I: Interpretation of XRD sample spectra for clay mineral identification comprising dry, heat-treated and glycol-solvated fractions of two representative samples. Reference spectra are included for mineral identification: clinochlore (PHILIPS et al. 1980), illite (DRITS et al. 2010), kaolinite (BISH & VON DREELE 1989), muscovite (RICHARD-SON & RICHARDSON 1982). Figure continued on next page.

268

Peak:	Position (2 to 22° 2Θ): (heat-treated / glycol / dry)	Interpretation:
A	- / 5.2 /-	Glycol-solvated smectite
B	6.3 / 6.3 / 6.3	Chlorite (clinochlore) + smectite
C	8.7 / 8.8 / 8.8	Muscovite / illite + heat-treated smectite
D	12.5/ 12.4 / 12.4	Chlorite (clinochlore) + kaolinite (?)
E	17.6 / 17.7 / 17.0	Muscovite / illite
F	- / 18.8 / 18.7	Chlorite (clinochlore) + chlorite heat-treated
G	19.8 / 19.8 / 18.9	Muscovite / illite
H	20.8 / 20.8 / 20.9	Quartz

Figure A 2.2, part II: The presence of smectite is indicated by a shift of the generally broad peak B (located at ~6.3° 2Θ) to position A (~5.23° 2Θ) in the glycol-solvated spectra, and the increase of peak C (8.78° 2Θ) after heat-treatment (MOORE & REY-NOLDS 1989). Peaks B and F (6.3 and 18.8° 2Θ) indicate clinochlore (chlorite) (MOORE & REYNOLDS 1989). Another typical clay mineral is kaolinite, which is difficult to identify in presence of chlorite. Heat treatment (550°C; 1h) causes de-hydroxylation and weakening of chlorite peaks similar to kaolinite. Hence, presence of kaolinite can in this case neither be excluded nor proved (MOORE & REYNOLDS 1989). The same problem arises in case of illite, which shares most of the peak positions with muscovite. While large plates of muscovite are directly visible in the sample material, the first order peak of illite at 19.8° 2Θ is barely visible.

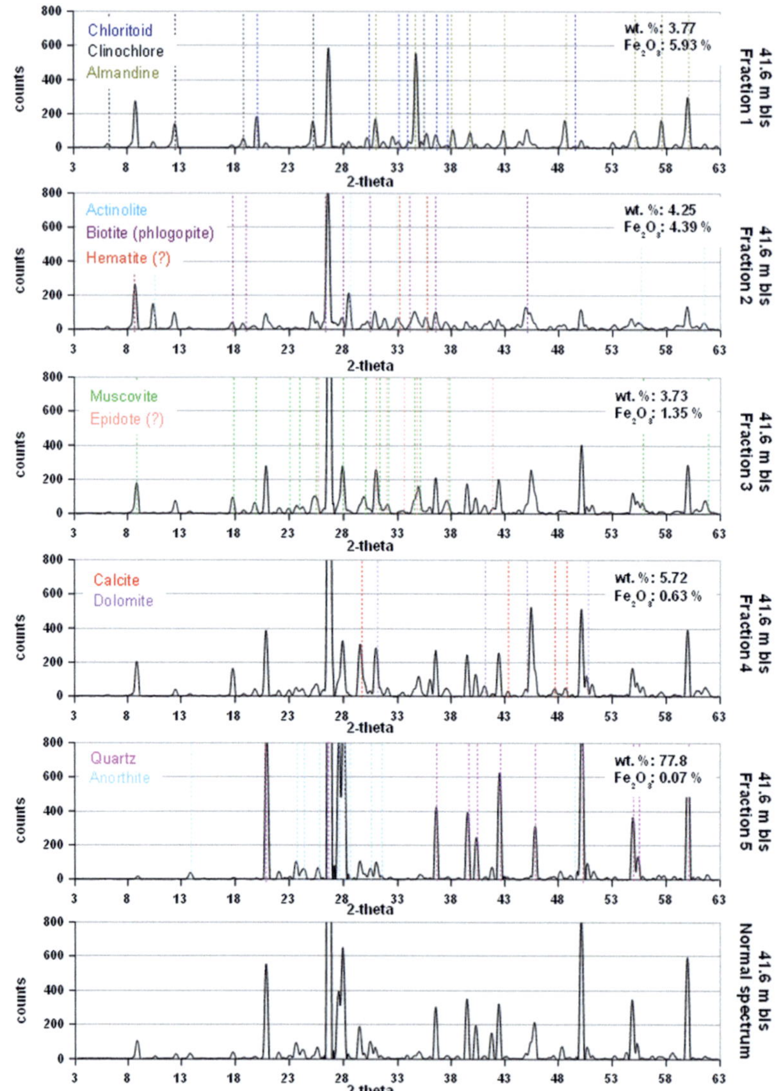

Figure A 2.3: Spectra of magnetic separated fractions for a sample from 41.6 m bls. Included are the fractions wt.% and the Fe content determined by EDX. Figure caption continued on next page.

Figure A 2.3, continued: Loss during fractionation: 4.37 wt.%; sample grain size comprised to 100 % the fraction <0.2 mm and >0.063mm. Dominating mineral phases were identified for each fraction. Reference spectra: actinolite (EVANS & YANG 1998), almandine (NOVAK & GIBBS 1971), anorthite (ANGEL 1988), calcite (GRAF 1961), chloritoid (HANSCOM 1980), clinochlore (PHILIPS et al. 1980), dolomite (STEINFINK & SANS 1959), epidote (DOLLASE 1971), hematite (BLAKE et al. 1966), muscovite (RICHARDSON & RICHARDSON 1982), phlogopite (STEINFINK 1962) and quartz (LEVIEN et al. 1980).

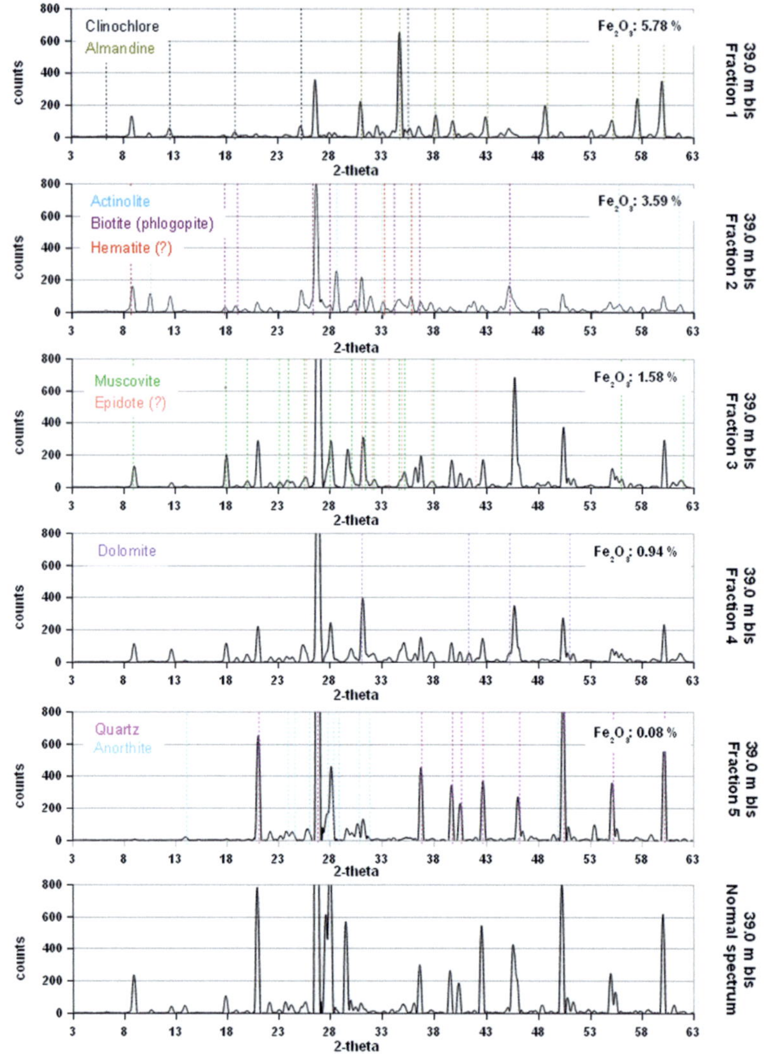

Figure A 2.4: Spectra for magnetic separated fractions for a sample from 39.0 m bls including respective Fe contents, which were determined by EDX. Grain size comprises to 100 % the fraction <0.2 mm and >0.063mm. For reference spectra, see Figure A 2.3.

Table A 2.3, part I: Literature values for OM characterisation (δ^{13}C measured against VPDB, δ^{15}N measured against air), based on different sample sources like mangrove trees and plant tissues. POM: particulate organic matter, M: marine, F: freshwater. Table continued on next page.

Ref.	Sample type	δ^{13}C (‰)			δ^{15}N (‰)	C/N (wt.%-ratio)		
		C_4	C_3	POM	C_3	C_4	C_3	POM
1	Summary	-15 to -10	-35 to -21	M: -20.3 to -22.4				
2	Plant tissues (marsh + bog), temperate zone		-25.0 to -31.3					
3	Summary			M: -19.0 to -24.5 F: -25 to -33				M,F: 4 to 10
4	Mangrove tissues		-27 to -30				18 to >50	
5	Summary	-10.8 to -14.1	-24.8 to -27.9	M: -19 to -25 F: -26.8 to -28.8		>20	>20	M,F: 3 to 10
6	Mangrove tissues		-25.9 to -29.1		-1.5 to +3.2		30 to 60	
7	Mangrove leaves and POM		-24.5 to -28.5	M: -20.5 to -21.5				
8	Summary			M: -20 to -24.5 F.: -24.5 to -29.0		5 to 70	20 to 78	M,F: 2 to 17
9	Summary							M: 4 to 18
10	Marsh vegetation	-12.3 to -12.8						

[1]HOEFS 2009, [2]HORNIBROOK et al. 2000, [3]LAMB et al. 2006, [4]MARCHAND et al. 2005, [5]MEYERS 1994, [6]MUZUKA & SHUNULA 2006, [7]RODELLI et al. 1984, [8]SARKA et al. 2009, [9]SHARP 2006, [10]STRIBLING & CORNWELL 1997.

Table A 2.3, part II: Additional literature values for OM characterisation including $\delta^{15}N$ values for C_4 plants and POM material. Sample sources comprise pollen and marine organic matter, including samples from the Bay of Bengal. POM: particulate organic matter, M: marine, F: freshwater.

Ref.	Sample type	$\delta^{13}C$ (‰)			$\delta^{15}N$ (‰)			C/N (wt.%-ratio)		
		C_4	C_3	POM	C_4	C_3	POM	C_4	C_3	POM
11	Pollen, temperate zone	-16 to -10	-21.5 to -29.5		+6.0 to +7.5	-3.8 to +13.8		13	5 to 55	
12	POM from Bay of Bengal						M: +3.43 to +4.29			
13	Marine TOC in sediment			M: -18.5 to -19.5						

[11]DESCOLAS-GROS & SCHÖLZEL 2007, [12]GAYE-HAAKE et al. 2005, [13]BLAIR & CARTER 1991.

Table A 2.4: Comparison of solid (s) and aqueous (a) phase compositions determined from groundwater (03/12/09) and corresponding sediments.

Well / sample depth (m bls)		Ca (ppm)	Mg (ppm)	Na (ppm)	K (ppm)	Mn (ppm)	Fe (ppm)	PO_4^{3-} (ppm)	As (ppb)	Zn (ppb)	Co (ppb)	Ni (ppb)	V (ppb)
A (12-21 / 13.0)	a*	86.5	20.6	18.6	2.49	0.77	4.34	2.38	49.1	7.48	4.33	0.52	0.08
	s**	17000	5480	9970	14400	409	15800	1320	2400	36800	4580	8180	33600
	r***	5.08×10^{-3}	3.77×10^{-3}	1.86×10^{-3}	1.73×10^{-4}	1.88×10^{-3}	2.75×10^{-4}	1.80×10^{-3}	2.03×10^{-2}	2.04×10^{-4}	9.45×10^{-4}	6.37×10^{-5}	2.48×10^{-6}
B (24-27 / 25.3)	a*	79.3	17.3	12.5	3.79	0.47	3.76	3.39	155	3.05	0.41	0.51	0.12
	s**	12700	2910	10100	13200	295	14200	1070	1800	25100	5180	10100	28400
	r***	6.23×10^{-3}	5.94×10^{-3}	1.24×10^{-3}	2.88×10^{-4}	1.61×10^{-3}	2.64×10^{-4}	3.18×10^{-3}	8.67×10^{-2}	1.22×10^{-4}	7.93×10^{-4}	5.00×10^{-5}	4.11×10^{-6}
C (30-33 / 30.5)	a*	69.5	17.4	11.4	6.82	0.41	5.57	1.12	135	4.50	0.29	0.36	0.02
	s**	20300	4720	10500	14200	212	10400	740	1500	22400	3270	6780	23700
	r***	3.43×10^{-3}	3.69×10^{-3}	1.09×10^{-3}	4.80×10^{-4}	1.92×10^{-3}	5.34×10^{-4}	1.51×10^{-3}	9.12×10^{-2}	2.01×10^{-4}	8.84×10^{-4}	5.26×10^{-5}	0.96×10^{-6}
D (36-39 / 38.3)	a*	67.6	15.8	10.8	3.72	0.42	2.86	2.18	132	5.79	0.37	0.37	0.08
	s**	13700	3040	9630	14600	473	17000	1220	2400	29800	6040	10600	34200
	r***	4.95×10^{-3}	5.20×10^{-3}	1.13×10^{-3}	2.55×10^{-4}	0.88×10^{-3}	1.68×10^{-4}	1.78×10^{-3}	5.61×10^{-2}	1.95×10^{-4}	6.04×10^{-4}	3.55×10^{-5}	2.20×10^{-6}
E (42-45 / 45.5)	a*	68.0	15.1	11.1	3.18	0.60	2.15	1.93	92.4	5.35	0.41	0.37	0.06
	s**	15100	4540	9460	14000	398	14700	1170	1900	27300	4910	9070	28600
	r***	4.50×10^{-3}	3.32×10^{-3}	1.18×10^{-3}	2.27×10^{-4}	1.50×10^{-3}	1.46×10^{-4}	1.65×10^{-3}	4.85×10^{-2}	1.96×10^{-4}	8.29×10^{-4}	4.13×10^{-5}	2.25×10^{-6}

*Groundwater concentration, representative results from 03/12/09

**Sediment content, determined from microwave acid digestions of representative samples that originate from the respective depth intervals of the well screenings

***Element ratio (aqueous phase / solid phase)

Table A 2.5, part I: Summary of sucrose experiment results, including initial baseline values. Results include concentrations of sucrose (sucr.), acetate (acet.), propionate (prop.) and butyrate (but.). Table continued on next page.

Date	T (°C)	EC (µS/cm)	pH	TA (mM)	O_2 (mg/L)	Sucr. (mg/L)	Acet. (mg/L)	Prop. (mg/L)	But. (mg/L)	TPC (cfu/mL)**	%-error (%)
					Well A (12-21 m bls)						
03/12	27.0	647	7.1	6.80	error	0	x	x	x	6.3×10^5	-1.17
06/12	27.8	612	7.1	7.40	error	8.20	0	0	0	x	-14.3
08/12	26.9	598	7.1	7.10	error	x	x	x	x	x	-13.4
10/12	27.2	638	7.1	7.50	error	3.30	0	0	0	2.9×10^7	-16.8
12/12	26.2	623	7.1	8.10	error	x	0	0	0	x	-14.2
14/12	27.2	621	7.0	7.95	error	9.80	0	0	0	1.0×10^8	-13.1
16/12	26.9	596	7.1	7.50	0.60	x	0	0	0	x	-13.5
18/12	26.9	561	7.2	8.10	0.08	11.5	x	x	x	1.1×10^8	-15.5
					Well B (24-27 m bls)						
03/12	27.0	623	7.2	7.90	0.53	0	x	x	x	4.1×10^5	-15.0
06/12	27.3	717	6.8	7.30	error	290	0	0	0	x	+0.20
08/12	27.0	1038	6.5	8.10	error	x	x	x	x	x	+23.4
10/12	26.8	1013	6.5	11.4	0.25	126	210	22.2	33.7	1.9×10^7	-4.93
12/12	26.3	1093	6.5	14.5	error	x	487	75.6	68.9	9.6×10^7	-5.37
14/12	28.2	1022	6.3	12.1	error	11.5	344	54.1	39.3	3.4×10^{10}	-0.62
16/12	27.1	907	6.6	11.1	0.49	x	322	48.9	34.5	x	+2.09
18/12	27.4	777	6.6	11.9	0.08	13.1	x	x	x	6.5×10^{10}	-7.36
					Well C (30-33 m bls)						
03/12	26.8	624	7.2	7.40	0.33	0	x	x	x	4.0×10^5	-16.3
06/12	26.9	643	7.1	7.70	0.21	35.0	0	0	0	x	-12.7
08/12	27.0	652	7.1	6.70	*0.18**	x	x	x	x	x	-3.85
10/12	26.6	928	6.8	12.1	0.14	468	210	0	0	3.3×10^7	-13.8
12/12	27.0	1067	6.5	12.4	error	x	431	25.2	27.2	6.0×10^7	-5.06
14/12	28.0	1052	6.5	11.6	error	67.0	295	15.6	12.8	1.9×10^{10}	-2.98
16/12	27.2	819	6.3	10.1	0.15	x	205	0	0	x	-7.51
18/12	27.8	729	6.7	10.7	0.12	13.1	x	x	x	1.3×10^{10}	-10.1

Outlier, replaced by mean of sample value before and after

**TPC: R2A-agar, 24 h, 37°C

x: not determined

Table A 2.5, part II. Table continued on next page.

Date	T (°C)	EC (µS/cm)	pH	TA (mM)	O_2 (mg/L)	Sucr. (mg/L)	Acet. (mg/L)	Prop. (mg/L)	But. (mg/L)	TPC (cfu/mL)**	%-error (%)
						Well D (36-39 m bls)					
03/12	26.5	604	7.2	7.60	0.24	0	x	x	x	3.5×10^5	-20.2
06/12	27.0	619	7.1	7.30	0.15	259	0	0	0	x	-15.0
08/12	26.9	669	7.1	7.50	*0.13**	x	x	x	x	x	-11.4
10/12	26.6	738	7.0	9.90	0.11	145	39.0	0	0	8.7×10^7	-19.5
12/12	27.0	817	6.9	9.00	*0.13**	x	128	0	0	5.9×10^7	-9.63
14/12	27.6	806	6.8	9.40	0.15	11.5	126	0	0	4.4×10^{10}	-10.7
16/12	27.4	791	6.9	9.50	0.14	x	153	0	0	x	-10.6
18/12	28.0	681	6.9	9.90	0.12	9.80	x	x	x	7.0×10^9	-12.9
						Well E (42-45 m bls)					
03/12	26.8	590	7.1	7.30	0.18	0	x	x	x	3.4×10^5	-18.5
06/12	26.9	611	6.7	6.40	0.13	873	0	0	0	x	+5.38
08/12	26.7	788	5.0	*6.65*	error	x	x	x	x	x	+3.70
10/12	*27.0*	1634	5.2	6.90	error	545	446	0	260	1.8×10^8	+41.4
12/12	27.2	1705	6.2	16.0	error	x	406	34.8	124	8.8×10^7	+9.31
14/12	27.2	1214	6.2	13.4	error	175	452	17.1	35.3	2.2×10^{10}	+1.43
16/12	27.5	966	6.4	10.4	0.22	x	334	15.6	24.0	x	-2.65
18/12	28.0	700	6.5	9.50	0.17	34.0	x	x	x	8.6×10^{10}	-11.0

Outlier, replaced by mean of sample value before and after

**TPC: R2A-agar, 24 h, 37°C

x: not determined

Table A 2.5, part III. Table continued on next page.

Date	Na	K	Ca	Mg	SIO_2	Cl^-	NO_3^-	SO_4^{2-}	PO_4^{3-}
					all (mg/L)				
Well A (12-21 m bls)									
03/12	18.6	2.49	86.5	20.6	34.8	7.85	<0.87	<0.85	2.38
06/12	14.4	2.42	72.3	16.7	29.7	5.78	<0.87	<0.85	2.23
08/12	14.1	2.18	71.0	16.6	29.3	6.18	<0.88	<0.85	5.42
10/12	14.5	2.07	68.4	16.4	28.0	7.37	<0.88	<0.85	5.22
12/12	17.1	2.21	78.6	18.6	31.4	7.67	<0.88	<0.85	5.60
14/12	17.0	2.18	79.6	18.7	31.3	7.83	<0.88	<0.85	5.77
16/12	16.2	2.15	73.8	17.8	29.0	7.91	<0.88	<0.85	5.33
18/12	17.0	2.27	77.3	18.2	30.1	8.40	<0.88	<0.85	5.72
Well B (24-27 m bls)									
03/12	12.5	3.79	79.3	17.3	31.5	2.65	<0.88	<0.85	3.39
06/12	15.0	3.75	96.3	21.4	34.9	4.57	<0.88	<0.85	2.44
08/12	16.2	4.24	179	31.1	41.3	3.23	<0.88	<0.85	8.60
10/12	14.3	3.86	139	24.8	33.8	2.61	<0.88	<0.85	6.61
12/12	15.0	4.56	183	31.4	38.6	2.99	<0.88	<0.85	7.21
14/12	14.2	4.41	166	28.5	36.4	4.24	<0.88	<0.85	7.11
16/12	13.9	4.41	162	27.9	35.4	2.75	<0.88	<0.85	6.95
18/12	13.3	4.24	145	25.3	33.5	3.13	<0.88	<0.85	6.52
Well C (30-33 m bls)									
03/12	11.4	6.82	69.5	17.4	28.5	2.43	<0.88	<0.85	1.12
06/12	14.8	3.95	75.1	18.8	27.9	5.65	<0.88	<0.85	4.18
08/12	14.8	4.01	78.7	19.4	30.3	*5.40*	<0.88	<0.85	8.48
10/12	15.6	4.50	119	26.2	32.5	5.14	<0.88	<0.85	8.31
12/12	15.7	5.38	151	31.0	34.6	5.00	<0.88	<0.85	7.02
14/12	14.5	5.48	149	28.6	32.2	4.43	<0.88	<0.85	5.97
16/12	13.1	4.89	117	21.7	29.1	3.69	<0.88	<0.85	5.54
18/12	13.1	5.21	122	21.6	29.5	3.78	<0.88	<0.85	5.42

Table A 2.5, part IV. Table continued on next page.

Date	Na	K	Ca	Mg	SIO_2	Cl^-	NO_3^-	SO_4^{2-}	PO_4^{3-}
					all (mg/L)				
Well D (36-39 m bls)									
03/12	10.8	3.72	67.6	15.8	26.6	2.64	<0.88	<0.85	2.18
06/12	13.8	3.05	69.0	16.8	23.6	5.41	<0.88	<0.85	2.24
08/12	14.1	3.19	76.8	18.5	27.2	5.04	<0.88	<0.85	7.45
10/12	14.5	3.22	84.5	20.4	28.1	4.76	<0.88	<0.85	7.24
12/12	14.0	3.45	95.1	22.2	28.6	4.77	<0.88	<0.85	7.37
14/12	14.2	3.49	99.7	23.1	29.4	5.00	<0.88	<0.85	7.17
16/12	14.0	3.47	101	23.1	29.2	4.61	<0.88	<0.85	6.31
18/12	13.7	3.55	101	23.1	30.1	4.64	<0.88	<0.85	6.13
Well E (42-45 m bls)									
03/12	11.1	3.18	68.0	15.1	25.5	2.59	<0.88	<0.85	1.93
06/12	14.3	3.09	93.3	21.2	27.7	3.60	<0.88	<0.85	5.30
08/12	14.2	2.59	97.5	19.0	26.8	3.80	<0.88	4.08	4.94
10/12	15.6	4.38	227	35.8	37.4	2.35	<0.88	5.32	8.17
12/12	14.2	4.74	260	36.7	37.3	2.06	<0.88	<0.85	2.65
14/12	12.6	4.21	194	25.4	33.6	2.49	<0.88	<0.85	5.13
16/12	11.8	3.70	139	19.2	30.4	2.58	<0.88	<0.85	4.83
18/12	11.3	3.29	108	15.6	28.2	2.41	<0.88	<0.85	4.48

Table A 2.5, part V. Table continued on next page.

Date	Fe (mg/L)	Fe(II) (%)	Mn (mg/L)	As (µg/L)	As(III) (%)	Zn (µg/L)	Ni (µg/L)	Co (µg/L)	V (µg/L)	Sr (µg/L)	Ba (µg/L)
					Well A (12-21 m bls)						
03/12	4.34	error	0.77	49.1	93.9	<u>7.48</u>	<u>0.52</u>	4.33	<u>0.08</u>	259	150
06/12	3.40	0.93	0.66	66.2	100	13.3	0.56	3.51	0.09	211	130
08/12	3.31	0.97	0.58	51.9	102	6.53	0.52	3.12	0.08	221	141
10/12	3.25	0.89	0.52	44.5	96.4	4.48	0.52	3.05	0.08	212	127
12/12	4.19	0.97	0.60	48.7	96.3	3.56	0.55	3.56	0.10	239	148
14/12	4.22	error	0.60	47.2	97.5	4.12	0.64	3.64	0.09	244	152
16/12	3.89	0.96	0.57	45.0	99.6	3.54	0.58	3.40	0.09	235	142
18/12	4.19	0.97	0.59	47.6	97.7	3.73	0.59	3.53	0.08	231	152
					Well B (24-27 m bls)						
03/12	3.76	error	0.47	155	95.8	3.05	0.51	0.41	0.12	267	140
06/12	8.58	0.86	1.01	110	74.8	41.5	3.86	3.45	*0.44**	311	224
08/12	29.7	0.99	2.49	217	93.3	70.1	2.34	7.48	0.77	558	399
10/12	20.2	0.97	1.67	159	100	28.8	1.56	3.25	0.39	439	332
12/12	29.8	0.99	2.30	204	94.7	55.5	2.00	4.95	0.23	583	423
14/12	28.0	error	2.15	199	97.3	42.0	1.87	4.60	0.16	507	389
16/12	26.6	0.98	1.98	203	95.8	130	1.76	4.26	0.14	498	366
18/12	24.2	0.96	1.77	190	96.0	28.8	1.58	3.82	0.13	445	345
					Well C (30-33 m bls)						
03/12	5.57	error	0.41	<u>135</u>	96.1	4.50	0.36	0.29	0.02	234	203
06/12	6.04	0.96	0.69	131	98.5	24.3	0.82	2.01	0.18	267	163
08/12	7.89	0.99	0.75	161	96.4	23.3	0.55	1.79	0.16	266	171
10/12	14.9	1.00	1.30	158	89.2	37.9	1.10	3.22	0.14	370	269
12/12	22.3	1.00	1.78	142	98.7	54.0	1.52	4.55	0.13	466	391
14/12	24.5	error	1.72	126	96.2	52.1	1.81	4.63	0.12	440	389
16/12	18.9	1.00	1.32	119	96.4	28.6	1.37	3.47	0.11	372	310
18/12	20.2	1.00	1.32	123	97.3	26.1	1.60	3.52	0.11	395	313

**Outlier*, replaced by mean of value before and after

<u>Outlier</u>, replaced by monitoring result (21/2/09)

Table A 2.5, part VI.

Date	Fe (mg/L)	Fe(II) (%)	Mn (mg/L)	As (µg/L)	As(III) (%)	Zn (µg/L)	Ni (µg/L)	Co (µg/L)	V (µg/L)	Sr (µg/L)	Ba (µg/L)
Well D (36-39 m bls)											
03/12	2.86	error	0.42	132	92.9	<u>5.79</u>	0.37	0.37	0.08	241	155
06/12	2.86	0.94	0.66	92	92.0	21.9	0.71	1.96	0.22	238	142
08/12	6.19	0.97	0.76	147	92.7	19.3	0.60	1.86	0.16	267	177
10/12	7.70	0.98	0.89	158	96.0	20.2	0.71	2.10	0.15	301	158
12/12	9.55	1.00	1.10	162	94.9	23.7	0.93	2.62	0.16	318	176
14/12	11.3	error	1.25	156	96.7	25.2	1.12	3.05	0.13	341	216
16/12	12.5	0.98	1.33	141	94.0	27.1	1.30	3.52	0.13	350	211
18/12	12.8	0.98	1.32	134	94.5	23.4	1.34	3.58	0.12	354	195
Well E (42-45 m bls)											
03/12	2.15	error	0.60	133	100	5.35	0.37	0.41	0.06	228	134
06/12	6.92	0.93	1.24	130	77.2	88.6	1.48	4.82	0.44	326	223
08/12	6.35	1.00	1.08	125	92.2	352	1.47	3.51	0.53	312	492
10/12	36.3	0.99	3.34	198	94.3	417	8.98	15.47	2.13	710	700
12/12	77.8	1.00	4.47	177	93.3	195	13.7	19.3	0.74	767	565
14/12	43.4	error	2.66	154	95.5	135	9.92	13.8	0.24	531	450
16/12	32.0	0.96	1.96	139	90.5	95.2	7.47	10.1	0.17	404	324
18/12	23.3	0.96	1.48	133	91.9	62.3	5.04	7.45	0.14	315	257

**Outlier*, replaced by mean of value before and after

<u>Outlier</u>, replaced by monitoring result (21/2/09)

Table A 2.6, part I: Biogeostratigraphic sediment interpretation for both study sites, which comprises lithology and chemistry of the sediment samples as well as isotopic values of embedded OM. HS: high As site, LS: low As site.

Facies	Depth (m bls)	Lithology	Sediment characterisation	Sequence interpretation
F1	HS: >39 LS: -	Clay	• No sample material available	*Sequence boundary LGM (?):* It is very difficult to interpret this layer without any sample material. Compared to the interpretation of SARKA et al. (2009), this clay can be considered as remains of top eroded marine shelf clay, which was surface exposed during the LGM (~20 ka BP). A significant sequence boundary developed to this time when the fluvial system eroded deep valleys into the clay (MC ARTHUR et al. 2008). Another possibility is that this layer consists of fluvial overbank clay and silt sediments, which were deposited after the LGM when the previously incised valleys were rapidly filled (see F3 below). This assumption is supported by analyses of drilling logs in the Nadia district done by MUKHERJEE et al. (2011) in order to model the local hydrological properties. Age: Late Pleistocene or early Holocene, maximum 20 ka bp
F2	HS: ~32 to 39.2 LS: ~29 to 45.5	Dark grey, weakly sorted sediments, dominated by fine sand medium sand with thin intrusion(s) of gravel, coarse sand and clay	HS: • TOC contents reach up to 0.53 wt.% and peak in the same depths as most trace elements and Ca. • Results of δ^{13}C, C/N and δ^{15}N analysis indicate a mixed signal of freshwater POM and C_3 plants, but interpretation of the data is difficult due to a potential loss of N. LS: • TOC too low (0.02-0.03 wt.%) for isotopic analysis, except for two samples from 31.8 and 33.1 m bls indicating marine POM signals, which are potentially influenced by terrestric plant matter.	*Lowstand system tract sequence following last glacial maximum (LGM):* Sediments from above the bottom clay layer are interpreted to originate from the time after the LGM (~20 ka bp). During this time, fluvial deposit filled incised valleys (GOODBRED et al. 2003). Embedded intrusions of mixed and poorly ordered muddy and gravelly (poor to medium rounded) sediments indicate high transport energies that allowed a rapid reallocation of older sediments. Gravels partly consist of secondary carbonate nodules, which have very likely developed during previous weathering. Increased TOC and mixed signals of freshwater POM and C_3 plants indicate that sediments of F2 originate from a freshwater swamp dominated environment. Age: Late Pleistocene or early Holocene, <20 ka bp

Table A 2.6, part II: Biogeostratigraphic sediment interpretation for both study sites, which comprises lithology and chemistry of the sediment samples as well as isotopic values of embedded OM. HS: high As site, LS: low As site.

Facies	Depth (m bls)	Lithology	Sediment characterisation	Sequence interpretation
F3A	HS: ~17 to ~32 LS: ~19 to ~29	Dark grey medium and fine sand	**HS:** • Sandy sediments are dominated by quartz and are depleted in most major and trace element contents, which exhibit throughout low variations. • TOC contents are constantly low, ranging between 0.02-0.03 wt.%. Only two samples hold enough TOC for isotopic analysis, which reflect a marine POM signal in 21.6 m bls and a freshwater POM signature in 24.2 m bls **LS:** • Sandy sediments are dominated by quartz and are depleted in most major and trace element contents, which exhibit throughout low variations. • TOC contents are too low (0.02-0.03 wt.%) for isotopic analysis.	*Transgressive system tract sequence:* Eustatic sea-level rise and presumably tectonic basin subsidence caused inland movement of the active shoreline (approx. 100 km), which was followed by mangrove-forests and swamps (SARKA et al. 2009). Fluvial and estuarine channel fill sands covered the incised valley fills of F2 allowing the development of a deltaic sedimentation environment with an upward increasing marine influence. A temperature increase beginning around ~9 ka BP intensified the seasonal monsoon precipitation, whereas sediment flux and sedimentation processes increased in the Bengal Basin (GOODBRED et al. 2003). Age: Holocene
F3B	HS: 3.35 to ~17 LS: 3.85 to ~19	Olive to dark grey silty fine sand	**HS:** • Fine grain content (<0.063 mm), TOC, major (K, Ca, TIC) and trace (Fe, Zn) element contents subsequently increase upwards, forming a distinctive boundary in ~17 m bls. • Enrichment of trace elements (Zr and Ce), which are characteristic for heavy minerals in ~13 m bls. • Organic matter δ^{13}C - C/N results plot in the field of marine POM, whereas eight δ^{15}N values are in a typical range of marine POM and three in range of C_3 vegetation. **LS:** • Fine grain content (<0.063 mm), TOC, main (K, Ca, TIC) and trace (Fe, Zn) element contents gradually increase upwards, forming a distinctive boundary in ~19 m bls. • Enrichment of trace elements (Zr and Ce), which are characteristic for heavy minerals. • Organic matter δ^{13}C - C/N results plot in the field of marine POM, whereas δ^{15}N values are in range of C_3 vegetation.	Subsequent deposition of huge amounts of fluvial and estuarine channel-fill sands changed the previously estuary character toward a more fluviatile influenced environment. Continuously increasing silt contents in upward direction indicate a change from an aggregating fluviatile-estuarine environment toward a low-relief floodplain system. Presumably, tidal-derived POM reflects a passing marine influence. Geochemical results reflect an increase in heavy minerals as indicated by Zr and Ce, which is decoupled from the silt content. This is likely linked to a temporary shift in the provenance. The redox boundaries located in about 6 m bls reflect persistent reducing conditions in respective aquifer parts. Above, the actual water table oscillates between dry and monsoon season, allowing relocation of labile elements into F4. Age: Holocene, <9 ka bp (?)

Table A 2.6, part III: Biogeostratigraphic sediment interpretation for both study sites, which comprises lithology and chemistry of the sediment samples as well as isotopic values of embedded OM. HS: high As site, LS: low As site.

Facies	Depth (m bls)	Lithology	Sediment characterisation	Sequence interpretation
F4	HS: 0 to 3.35 LS: 0 to 3.85	Brown silt and clay	HS: • Enrichment of Fe, Mn and typically co-precipitating trace elements (As, Ni, Cu, Zn), while Ca and TIC are depleted. Maximum Fe (9.62 wt.% Fe_2O_3) and As (122 mg kg^{-1}) contents appear between 2.65 and 3.35 m bls. • TOC contents of up to 0.38 wt.% are clearly increased compared to sediments below. • $\delta^{13}C$ - C/N and $\delta^{15}N$ values indicate a shift from marine POM toward a mixture of POM with C_4 vegetation. • Clay minerals are dominated by smectite and kaolinite. • Clinker fragments reflect an anthropogenic influence down to ~2.00 m bls. LS: • Between 1.90-3.20 m bls., enrichment of TOC, Fe, Mn and typically co-precipitating trace elements (As, Ni, Cu, Zn), while Ca and TIC are depleted. Maximum Fe (8.06 wt.% Fe_2O_3) and Mn (0.09 wt.% MnO) contents appear around 2.55 m bls. • $\delta^{13}C$ - C/N and $\delta^{15}N$ values indicate a shift from marine POM toward a mixture of POM with C_4 vegetation. • Clay minerals are dominated by smectite and kaolinite. • Clinker fragments and compaction of the sediments within top 0.60 cm	*Delta-floodplain progradation sequence:* The top silty and clayey sediments originate from the present floodplain and are interpreted as latest fluvial channel-overbank deposit of the Hooghly River (MC ARTHUR et al. 2008). OM signatures reflect an increasing influence of C_4 plants, which are the principal type of today's floodplain vegetation (SARKA et al. 2009). Capillary rise of reduced groundwater into the oxic unsaturated zone causes precipitation of Fe and other redox-sensitive trace elements. Clinker fragments indicate that upper sediments have been disturbed during construction of the courtyard. Age: Late Holocene

A.3   THE HIGH ARSENIC STUDY SITE

Table A 3.1: Summary of sediment compositions, comprising all samples from 0 to 39.2 m bls (n: 61).

Value	K_2O (wt.%)	CaO (wt.%)	TiO_2 (wt.%)	MnO (wt.%)	Fe_2O_3 (wt.%)	As (mg/kg)	Ni (mg/kg)	Cu (mg/kg)	Zn (mg/kg)
Minimum	1.30	1.09	0.18	0.02	1.36	2.3	21.6	8.58	19.3
25 % Q.	1.81	1.99	0.28	0.03	2.08	3.0	24.0	9.72	30.2
Median	2.09	2.89	0.36	0.05	2.62	3.6	25.9	11.3	36.7
75 % Q.	2.64	3.80	0.50	0.05	3.52	4.5	29.5	14.6	47.5
Maximum	3.91	6.16	0.85	0.11	9.62	122	112	70.4	117
Average	2.27	2.92	0.41	0.05	3.14	6.6	30.1	16.2	42.5
r_{As-}	0.47	-0.20	0.46	0.52	0.63	-	0.48	0.70	0.62
r_{Fe-}	0.83	0.13	0.91	0.86	-	0.63	0.85	0.96	0.99

Value	Ba (mg/kg)	Sr (mg/kg)	Ce (mg/kg)	Y (mg/kg)	La (mg/kg)	TS (mg/kg)	Fine fraction* (wt.%)	TOC (wt.%)	TIC (wt.%)
Minimum	245	76.7	31.3	10.2	16.2	<25	0.32	0.02	0.03
25 % Q.	311	87.2	52.4	16.3	25.9	<25	4.46	0.03	0.26
Median	337	101	66.8	20.5	31.3	<25	11.8	0.04	0.41
75 % Q.	378	120	82.6	26.1	38.5	46.2	22.4	0.09	0.58
Maximum	566	238	214	54.0	91.9	256	92.8	0.53	1.00
Average	356	109	71.1	22.1	33.8	34.7	19.9	0.10	0.43
r_{As-}	0.46	-0.16	0.06	0.23	0.05	0.02	0.52	0.38	-0.23
r_{Fe-}	0.90	-0.13	0.30	0.57	0.29	0.40	0.87	0.67	0.10

*Fraction <0.063 mm, comprising silt and clay

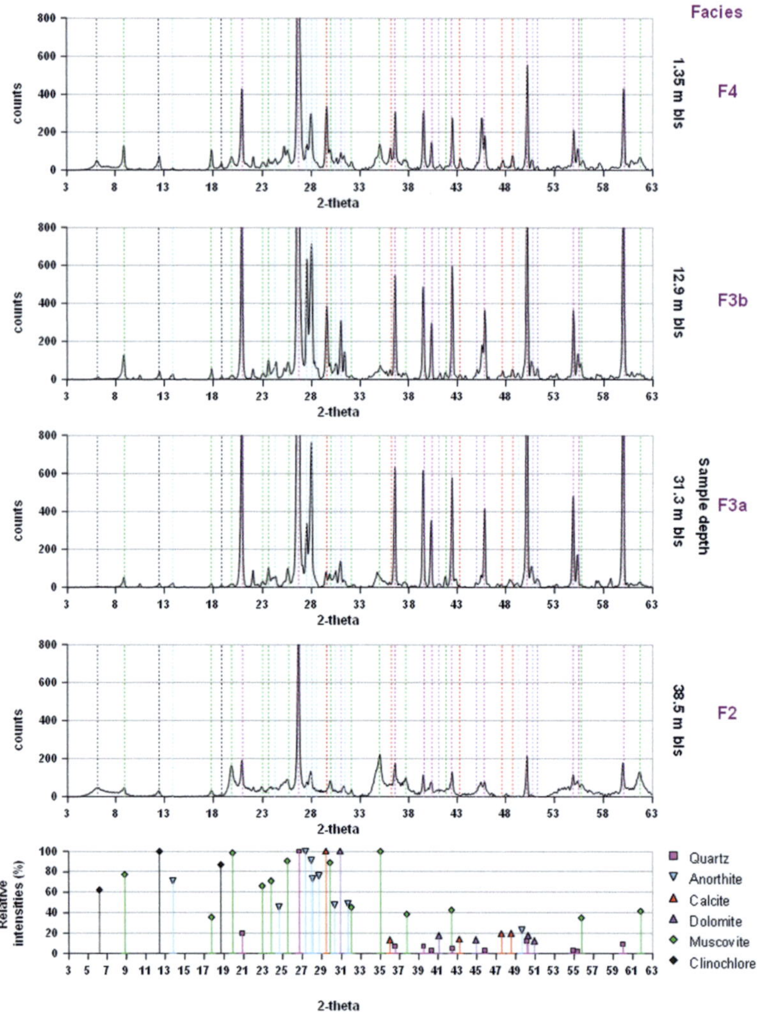

Figure A 3.1: Sample XRD spectra from different depths at the high As site. Reference spectra of identified mineral peaks see APPENDIX II, Figure A 2.1

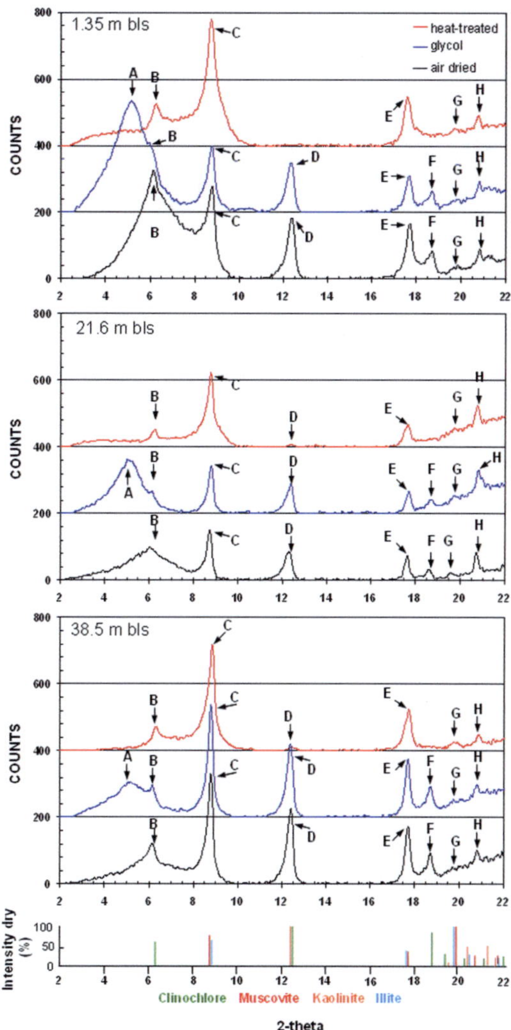

Figure A 3.2: Identification of clay minerals including dry, heat-treated and glycol-solvated subsamples. For interpretation of peak positions and reference spectra see APPENDIX II, Figure A 2.2.

Table A 3.2: Distribution of As, Fe and Mn in SEP fractions (I-VII) of representative samples (fractions described in Table A 1.4). GXR-2 reference values (mg/kg): As: 25.0; Fe: 19,000; Mn: 1,007.

Depth (m bls)	As (mg/kg)							
	I	II	III	IV	V	VI	VII	Σ
2.65	0.00	0.19	0.30	0.77	1.87	0.00	0.57	3.70
3.35	7.68	74.0	4.91	26.65	7.94	0.00	0.83	122
12.4 A	0.00	2.54	0.20	0.77	0.68	0.00	0.70	4.89
12.4 B	0.00	1.36	0.21	0.97	0.40	0.00	1.79	4.73
23.5	0.00	0.00	0.10	0.07	0.22	0.00	0.28	0.67
29.4	0.00	0.91	0.17	0.25	0.19	0.00	0.32	1.84
30.7	0.00	1.77	0.12	0.36	0.25	0.00	0.27	2.77
38.5	0.13	4.41	0.48	0.35	0.85	0.00	0.75	6.96
GXR-2	1.21	21.4	2.34	2.11	1.38	0.00	0.35	28.8

Depth (m bls)	Fe (mg/kg)							
	I	II	III	IV	V	VI	VII	Σ
2.65	421	5123	4745	3125	4587	26 162	132	44 296
3.35	1853	4530	2990	6729	16 171	27 242	23.1	59 537
12.4 A	35.9	1754	2887	1391	2536	11 204	385	20 191
12.4 B	30.2	1187	2112	2479	1688	11 618	512	19 627
23.5	19.2	1197	1658	820	975	4973	591	10 233
29.4	13.4	1526	2434	1465	1829	7430	218	14 915
30.7	16.3	1766	2153	1482	2256	8443	817	16 933
38.5	155	4765	8776	1709	3035	14 805	157	33 402
GXR-2	169	3175	6362	1698	6043	5043	95.5	22 586

Depth (m bls)	Mn (mg/kg)							
	I	II	III	IV	V	VI	VII	Σ
2.65	26.3	281	58.9	40.8	27.0	293	6.15	733
3.35	9.88	254	205	449	84.0	200	0.31	1202
12.4 A	13.2	68.5	41.7	0.00	14.7	216	29.3	383
12.4 B	13.6	63.2	33.7	11.9	7.73	269	40.1	440
23.5	14.0	35.1	24.2	14.6	6.17	77.2	39.9	211
29.4	12.9	78.1	71.8	95.8	14.9	112	15.7	402
30.7	9.56	79.4	57.5	110	16.2	127	56.2	456
38.5	80.4	272	126	22.9	12.0	157	5.63	676
GXR-2	149	886	72.1	36.6	28.8	76.2	1.55	1250

*SEP Fractions:

I: Weakly bound As; II: Strongly adsorbed As; III: Incorporated in AVS, carbonates, Mn-oxides, very amorphous Fe- (oxyhydr)oxides; IV: Incorporated in amorphous Fe-(oxyhydr)oxides; V: Incorporated in crystalline Fe-(oxyhydr)oxides; VI: Associated with silicates.

Table A 3.3, part I: Summary of the groundwater abstraction experiment results, including initial baselines (03/12/09). No outliers were removed. Table continued on next page.

Well	Date	T (°C)	Odour*	EC (µS/cm)	pH	E_H (mV)	TA (mM)	$\delta^{18}O$ (‰ VSMOW)	δ^2H (‰ VSMOW)
A	03/12	26.0	none	973	6.9	x	9.60	-2.95	-23.3
	08/12	26.4	H_2S (s)	896	7.1	x	8.30	-3.11	-24.1
	11/12	26.7	H_2S (w)	777	7.1	-52	9.40	-3.84	-26.6
	13/12	26.5	H_2S (m)	761	7.0	-52	9.80	-4.17	-26.6
	14/12	26.1	H_2S (w)	712	7.1	error	9.40	-3.79	-26.2
	17/12	26.6	H_2S (m)	683	7.2	x	9.20	-3.70	-25.3
B	03/12	26.6	none	1040	6.9	x	10.0	-3.07	-23.6
	08/12	26.3	none	962	6.9	x	8.80	-3.12	-19.9
	11/12	26.0	H_2S (w)	796	7.0	-67	9.70	-4.06	-24.8
	13/12	26.5	H_2S (m)	679	7.1	-65	8.00	-4.01	-31.7
	14/12	26.4	H_2S (w)	676	7.1	-56	8.20	-4.28	-29.1
	17/12	26.6	H_2S (s)	630	7.2	x	8.50	-4.15	-27.8
C	03/12	27.2	none	718	7.2	x	7.90	-3.92	-28.3
	08/12	26.3	none	670	7.3	x	7.20	-3.94	-27.1
	11/12	26.7	H_2S (w)	693	7.2	-43	8.80	-4.13	-25.4
	13/12	26.7	none	664	7.2	-44	8.30	-3.95	-28.2
	14/12	26.2	none	660	7.2	-42	7.80	-4.18	-28.7
	17/12	26.6	H_2S (w)	636	7.2	x	8.20	-4.17	-27.1
D	03/12	26.9	none	781	7.1	x	8.20	-4.17	-30.9
	08/12	26.8	none	764	7.2	x	9.00	-4.12	-29.1
	11/12	26.9	H_2S (w)	787	7.1	-72	10.9	-4.42	-28.7
	13/12	26.8	H_2S (m)	758	7.1	-57	10.1	-4.00	-27.4
	14/12	26.3	H_2S (w)	737	7.2	-48	9.60	-4.29	-30.7
	17/12	26.9	H_2S (w)	704	7.2	x	9.60	-4.31	-28.0
E	03/12	26.9	none	745	7.0	x	8.40	-3.81	-27.7
	08/12	26.9	none	715	7.1	x	7.80	-3.84	-27.5
	11/12	26.6	H_2S (m)	744	7.1	-58	9.70	-4.22	-27.2
	13/12	26.7	H_2S (m)	716	7.1	-45	9.00	-3.86	-26.9
	14/12	26.2	H_2S (w)	687	7.1	-47	9.50	-4.07	-29.3
	17/12	26.8	H_2S (m)	654	7.3	x	8.60	-4.24	-27.7

*Odour intensity: weak (w), medium (m), strong (s)

x: not determined

Table A 3.3, part II. Table continued on next page.

Well	Date	Na	Mg	K	Ca	SiO$_2$	Cl$^-$	NO$_3^-$	SO$_4^{2-}$	PO$_4^{3-}$	DOC
						all (mg/L)					
A	03/12	29.4	7.77	95.5	0.92	98.0	85.1	<0.88	7.78	2.61	3.02
	08/12	28.0	6.93	89.4	0.71	170	100	<0.88	8.34	6.54	x
	11/12	22.4	4.67	x	0.43	300	98.5	<0.88	1.72	7.16	x
	13/12	22.0	4.88	x	0.38	335	96.2	<0.88	1.51	11.8	x
	14/12	21.3	5.14	97.5	0.40	326	98.1	<0.88	1.63	8.05	x
	17/12	20.5	5.00	x	0.36	343	98.4	<0.88	1.34	7.63	8.31
B	03/12	27.8	8.96	98.6	0.74	100	99.0	<0.88	10.7	2.99	4.52
	08/12	24.6	8.61	79.8	0.56	150	97.6	<0.88	9.51	6.83	x
	11/12	19.8	6.35	x	0.35	254	95.1	<0.88	3.51	6.97	x
	13/12	17.8	5.12	x	0.27	330	99.0	<0.88	1.32	7.76	x
	14/12	17.5	4.64	93.6	0.28	312	98.3	<0.88	1.43	7.25	x
	17/12	17.4	4.61	x	0.24	349	95.5	<0.88	<0.85	6.98	6.68
C	03/12	18.3	1.09	89.8	0.85	296	94.4	<0.88	<0.85	2.53	2.55
	08/12	17.9	1.47	error	0.85	335	91.9	<0.88	<0.85	6.66	x
	11/12	17.6	1.09	x	0.80	301	82.7	<0.88	<0.85	6.26	x
	13/12	17.1	1.39	x	0.79	326	96.5	<0.88	<0.85	6.06	x
	14/12	16.6	1.48	100	0.77	325	97.1	<0.88	<0.85	5.95	x
	17/12	17.0	1.77	x	0.77	334	98.0	<0.88	<0.85	6.48	5.60
D	03/12	19.9	4.76	99.1	0.37	262	95.9	0.97	<0.85	3.08	1.53
	08/12	21.9	4.43	81.4	0.43	245	97.0	<0.88	<0.85	6.54	x
	11/12	21.7	4.97	x	0.40	245	95.7	<0.88	<0.85	7.33	x
	13/12	21.0	4.78	x	0.38	251	97.7	<0.88	<0.85	5.76	x
	14/12	21.3	4.81	100	0.40	246	83.0	<0.88	<0.85	6.14	x
	17/12	20.8	5.04	x	0.38	246	96.1	<0.88	<0.85	5.94	6.17
E	03/12	14.1	4.80	98.2	0.47	158	99.2	<0.88	<0.85	2.84	1.32
	08/12	15.5	4.65	85.2	0.52	158	97.8	<0.88	<0.85	5.84	x
	11/12	16.0	4.86	x	0.51	159	100	<0.88	1.07	6.20	x
	13/12	16.9	4.96	x	0.50	165	95.1	<0.88	1.62	5.73	x
	14/12	15.8	4.71	98.0	0.50	160	99.0	<0.88	1.25	5.82	x
	17/12	16.2	4.99	x	0.46	164	94.3	<0.88	1.75	5.61	6.04

x: not determined

Table A 3.3, part III.

Well	Date	Fe (mg/L)	Fe(II) (%)	Mn (mg/L)	As (µg/L)	As(III) (%)	Rb	Li	U	Mo	%-error (%)
A	03/12	7.77	95.5	0.92	98.0	85.1	1.45	2.15	0.08	1.21	-1.63
	08/12	6.93	89.4	0.71	170	100	2.22	2.41	0.04	0.84	+1.23
	11/12	4.67	x	0.43	300	98.5	2.65	2.26	0.02	0.94	-8.42
	13/12	4.88	x	0.38	335	96.2	3.29	2.22	0.01	2.27	-12.0
	14/12	5.14	97.5	0.40	326	98.1	2.68	2.09	0.03	1.00	-10.8
	17/12	5.00	x	0.36	343	98.4	2.69	1.99	0.02	0.92	-11.2
B	03/12	8.96	98.6	0.74	100	99.0	4.56	2.76	0.08	0.86	-4.76
	08/12	8.61	79.8	0.56	150	97.6	3.72	3.13	0.07	0.54	+1.07
	11/12	6.35	x	0.35	254	95.1	3.15	2.65	0.06	1.01	-9.08
	13/12	5.12	x	0.27	330	99.0	2.82	2.50	0.03	1.43	-3.50
	14/12	4.64	93.6	0.28	312	98.3	2.75	2.23	0.05	5.07	-6.37
	17/12	4.61	x	0.24	349	95.5	2.72	2.24	0.04	1.45	-9.66
C	03/12	1.09	89.8	0.85	296	94.4	1.40	3.06	0.58	1.97	-6.32
	08/12	1.47	error	0.85	335	91.9	1.38	2.78	0.59	1.83	-2.41
	11/12	1.09	x	0.80	301	82.7	1.41	2.45	0.57	7.25	-13.3
	13/12	1.39	x	0.79	326	96.5	1.45	2.16	0.55	1.83	-9.79
	14/12	1.48	100	0.77	325	97.1	1.35	1.99	0.49	1.74	-9.02
	17/12	1.77	x	0.77	334	98.0	1.40	2.08	0.52	1.73	-10.6
D	03/12	4.76	99.1	0.37	262	95.9	2.00	2.32	0.16	1.23	-1.59
	08/12	4.43	81.4	0.43	245	97.0	1.94	1.93	0.22	1.03	-5.50
	11/12	4.97	x	0.40	245	95.7	1.98	1.89	0.15	1.00	-14.1
	13/12	4.78	x	0.38	251	97.7	1.89	1.87	0.12	0.86	-12.7
	14/12	4.81	100	0.40	246	83.0	1.89	1.88	0.15	0.93	-9.47
	17/12	5.04	x	0.38	246	96.1	1.89	1.89	0.12	0.86	-11.4
E	03/12	4.80	98.2	0.47	158	99.2	1.92	1.70	0.16	1.57	-15.2
	08/12	4.65	85.2	0.52	158	97.8	1.95	1.66	0.18	1.61	-6.87
	11/12	4.86	x	0.51	159	100	1.95	1.75	0.19	1.67	-14.6
	13/12	4.96	x	0.50	165	95.1	1.98	1.79	0.17	1.49	-10.4
	14/12	4.71	98.0	0.50	160	99.0	1.93	1.69	0.19	1.58	-14.1
	17/12	4.99	x	0.46	164	94.3	1.94	1.77	0.17	1.61	-10.0

x: not determined

Table A 3.4: Comparison of solid phase compositions and aqueous phase concentrations in different depths.

Well/ sample depth (m bls)	Source	Ca (ppm)	Mg (ppm)	Na (ppm)	K (ppm)	Mn (ppm)	Fe (ppm)	PO_4^{3-} (ppm)	Ba (ppm)	As (ppb)	U (ppb)	TOC / DOC (ppm)
A 12-21 / 12.4	Water*	134	28.4	29.36	9.73	0.92	7.77	2.61	0.33	98.0	0.08	3.02
	Sediment**	22 000	5630	9920	13 900	490	20 000	2000	299	5600	5170	380
	Ratio***	6.09×10^{-3}	5.04×10^{-3}	2.96×10^{-3}	7.01×10^{-4}	1.87×10^{-3}	3.88×10^{-4}	1.33×10^{-3}	1.10×10^{-3}	1.75×10^{-2}	1.54×10^{-5}	7.97×10^{-3}
B 22-25 / 23.5	Water	140	27.7	27.8	9.70	0.74	8.96	2.99	0.43	100	0.08	4.52
	Sediment	10 100	3650	10 530	14 700	256	10 300	706	301	2000	1370	208
	Ratio	14.2×10^{-3}	7.59×10^{-3}	2.64×10^{-3}	6.60×10^{-4}	2.90×10^{-3}	8.67×10^{-4}	4.24×10^{-3}	1.42×10^{-3}	5.07×10^{-2}	5.95×10^{-5}	21.7×10^{-3}
C 26-29 / 29.4	Water	93.9	23.3	18.3	2.73	0.85	1.09	2.53	0.17	296	0.58	2.55
	Sediment	12 900	3900	10 100	14 700	388	14 600	961	310	2500	1520	333
	Ratio	7.29×10^{-3}	5.98×10^{-3}	1.82×10^{-3}	1.85×10^{-4}	2.20×10^{-3}	0.75×10^{-4}	2.63×10^{-3}	0.56×10^{-3}	12.1×10^{-2}	38.1×10^{-5}	7.64×10^{-3}
D 30-33 / 30.6	Water	109	24.7	19.9	2.72	0.37	4.76	3.07	0.26	262	0.16	1.53
	Sediment	13 100	4020	10 600	15 200	452	15 700	1150	330	3300	1910	286
	Ratio	8.34×10^{-3}	6.14×10^{-3}	1.88×10^{-3}	1.79×10^{-4}	0.83×10^{-3}	3.04×10^{-4}	2.67×10^{-3}	0.78×10^{-3}	7.93×10^{-2}	8.45×10^{-5}	5.35×10^{-3}
E 34-37 / 38.5	Water	92.0	22.7	14.1	2.88	0.47	4.80	2.84	0.23	158	0.16	1.32
	Sediment	41 200	16 900	8840	26 600	691	36 800	1600	555	9700	3160	5320
	Ratio	2.23×10^{-3}	1.34×10^{-3}	1.60×10^{-3}	1.09×10^{-4}	0.69×10^{-3}	1.31×10^{-4}	1.77×10^{-3}	0.41×10^{-3}	1.64×10^{-2}	5.18×10^{-5}	0.25×10^{-3}

*Groundwater concentrations, representative results from 03/12/09

**Sediment contents, determined by microwave acid digestion from representative samples according to the well's screening positions

***Element ratios aqueous phase / solid phase

TOC vs TN

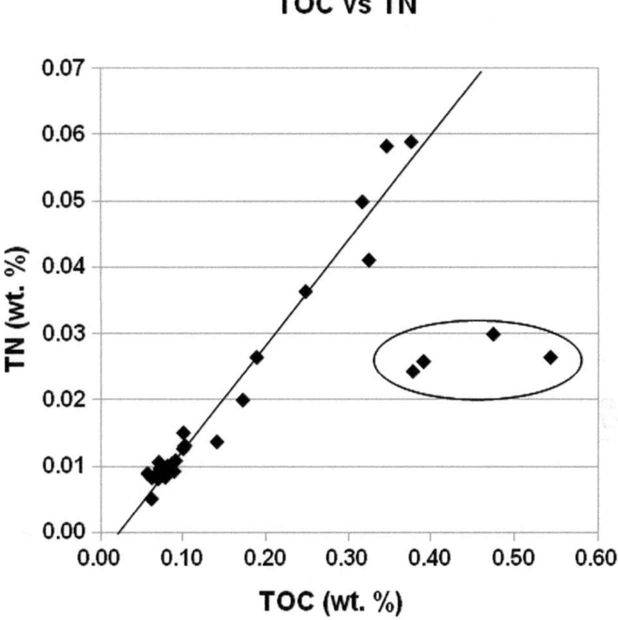

Figure A 3.3: Comparison of TN and TOC contents of the sediments, which display a group of outliers that originate from facies F2. Regression line without highlighted outlier group (R^2: 0.97, n: 27). This decoupling in the four samples is attributed to diagenetic alteration effects.

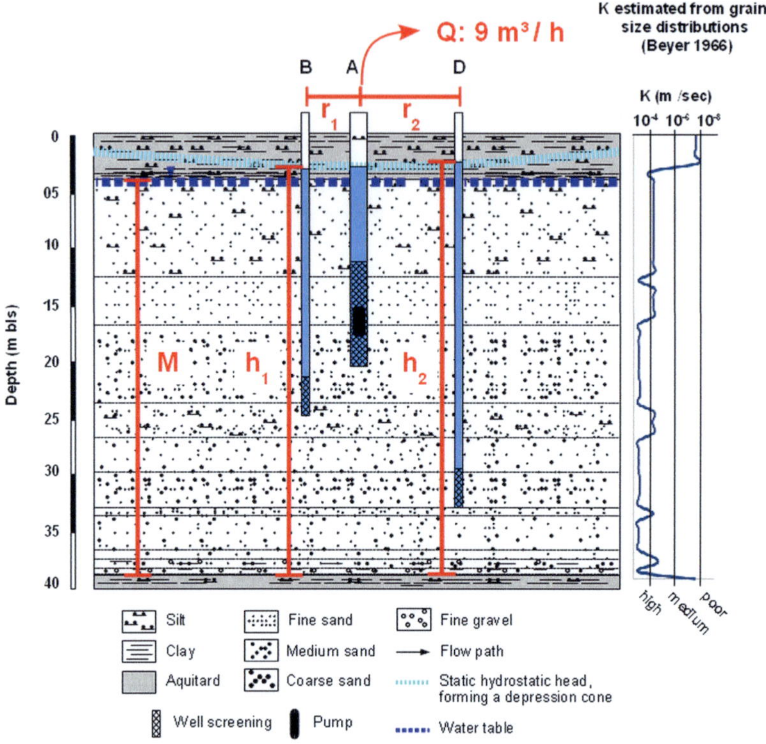

Figure A 3.4: Calculation of the K value with the Dupuit-Thieme equation and comparison to grain size estimated K values. Since most assumptions are not directly met, the calculated K is considered a rough estimate, but is in relatively good agreement with the grain size estimated K values and sufficient for considerations in the context of this study.

Calculation of hydraulic conductivity (K value) with the Dupuit-Thieme equation (ENTENMANN 2006):

$$Q = K \times 2M \times \left(\frac{(h_2 - h_1)}{\ln \dfrac{r_2}{r_1}} \right) \qquad (A1)$$

Q: 9 m^3 h^{-1} (delivery rate)

M: 38.85 m (thickness of water column, 39.2 m - 3.35 m);

h_1: 38.92 m (distance hydrostatic head in well B to bottom aquitard

= 39.2 m – 0.28 m);

h_2: 39 m (distance hydrostatic head in well D to bottom aquitard

= 39.2 m – 0.20 m);

r_1: 1.96 m (distance well B to pumping well)

r_2: 4.85 m (distance well D to pumping well)

$\rightarrow$ K = 1.26 x 10^{-4} m sec^{-1}

Assumptions:

1) Aquifer is confined;

2) Flow conditions are quasi-stationary (steady state conditions);

3) Aquifer is homogenous and has an unlimited extension;

4) The hydrostatic head has no gradients;

5) The well screens cover the full vertical extension of the aquifer;

6) Constant delivery rate.

In der Reihe „Karlsruher Geochemische Hefte"
(ISSN 0943-8599) sind erschienen:

Band 1: U. Kramar (1993)
Methoden zur Interpretation von Daten der geochemischen
Bachsedimentprospektion am Beispiel der Sierra de San Carlos/ Tamaulipas
Mexiko

Band 2: Z. Berner (1993)
S-Isotopengeochemie in der KTB - Vorbohrung und Beziehungen zu den
Spuren-elementmustern der Pyrite

Band 3: J.-D. Eckhardt (1993)
Geochemische Untersuchungen an jungen Sedimenten von der Galapagos-
Mikroplatte: Hydrothermale und stratigraphisch signifikante Signale

Band 4: B. Bergfeldt (1994)
Lösungs- und Austauschprozesse in der ungesättigten Bodenwasserzone
und Auswirkungen auf das Grundwasser

Band 5: M. Hodel (1994)
Untersuchungen zur Festlegung und Mobilisierung der Elemente As, Cd, Ni
und Pb an ausgewählten Festphasen unter besonderer Berücksichtigung des
Einflusses von Huminstoffen.

Band 6: T. Bergfeldt (1995)
Untersuchungen der Arsen- und Schwermetallmobilität in Bergbauhalden
und kontaminierten Böden im Gebiet des Mittleren Schwarzwaldes

Band 7: M. Manz (1995)
Umweltbelastungen durch Arsen und Schwermetalle in Böden, Halden,
Pflanzen und Schlacken ehemaliger Bergbaugebiete des Mittleren und
Südlichen Schwarzwaldes.

Band 8: J. Ritter (1995)
Genese der Mineralisation Herrmanngang im Albtalgranit (SE-Schwarzwald)
und Wechselwirkungen mit dem Nebengestein.

Band 9: J. Castro (1995)
Umweltauswirkungen des Bergbaus im semiariden Gebiet von Santa Maria de la Paz, Mexiko.

Band 10: T. Rüde (1996)
Beiträge zur Geochemie des Arsens.

Band 11: J. Schäfer (1998)
Einträge und Kontaminationspfade Kfz-emittierter Platin-Gruppen-Elemente (PGE) in verschiedenen Umweltkompartimenten.

Band 12: M. A. Leosson (1999)
A Contribution to the Sulphur Isotope Geochemistry of the Upper Continental Crust: The KTB Main Hole - A Case Study

Band 13: B. A. Kappes (2000)
Mobilisierbarkeit von Schwermetallen und Arsen durch saure Grubenabwässer in Bergbaualtlasten der Ag-Pb-Zn-Lagerstätte in Wiesloch

Band 14: H. Philipp (2000)
The behaviour of platinum-group elements in petrogenetic processes: A case study from the seaward-dipping reflector sequence (SDRS), Southeast Greenland volcanic rifted margin

Band 15: E. Walpersdorf (2000)
Nähr- und Spurenelementdynamik im Sediment/Wasser-Kontaktbereich nach einer Seekreideaufspülung – Pilotstudie Arendsee –

Band 16: R. H. Kölbl (2000)
Models of hydrothermal plumes by submarine diffuse venting in a coastal area: A case study for Milos, South Aegean Volcanic Arc, Greece

Band 17: U. Heiser (2000)
Calcium-rich Rhodochrosite layers in Sediments of the Gotland Deep, Baltic Sea, as Indicators for Seawater Inflow

In der Fortsetzungsreihe „Karlsruher Mineralogische und
Geochemische Hefte" (ISSN 1618-2677) sind bisher erschienen:

Band 18: S. Norra (2001)
Umweltgeochemische Signale urbaner Systeme am Beispiel von Böden,
Pflanzen, und Stäuben in Karlsruhe

Band 19: M. von Wagner, Alexander Salichow, Doris Stüben (2001)
Geochemische Reinigung kleiner Fließgewässer mit Mangankiesen, einem
Abfallsprodukt aus Wasserwerken (GReiFMan)

Band 20: U. Berg (2002)
Die Kalzitapplikation als Restaurierungsmaßnahme für eutrophe Seen –
ihre Optimierung und Bewertung

Band 21: Ch. Menzel (2002)
Bestimmung und Verteilung aquatischer PGE-Spezies in urbanen Systemen

Band 22: P. Graf (2002)
Meta-Kaolinit als latent-hydraulische Komponente in Kalkmörtel

Band 23: D. Buqezi-Ahmeti (2003)
Die Fluidgehalte der Mantel-Xenolithe des Alkaligesteins-Komplexes
der Persani-Berge, Ostkarpaten, Rumänien: Untersuchungen an Fluid-
Einschlüssen

Band 24: B. Scheibner (2003)
Das geochemische Verhalten der Platingruppenelemente bei der Entstehung
und Differenzierung der Magmen der Kerguelen-Flutbasaltprovinz (Indischer
Ozean)

**Ab Band 25 erscheinen die Karlsruher Mineralogischen und
Geochemischen Hefte bei KIT Scientific Publishing online unter der
Internetadresse http://www.ksp.kit.edu
Auf Wunsch sind bei KIT Scientific Publishing gedruckte Exemplare
erhältlich („print on demand").**

Band 25: A. Stögbauer (2005)
Schwefelisotopenfraktionierung in abwasserbelasteten Sedimenten
- Biogeochemische Umsetzungen und deren Auswirkung auf den
Schwermetallhaushalt

Band 26: X. Xie (2005)
Assessment of an ultramicroelectrode array (UMEA) sensor for the
determination of trace concentrations of heavy metals in water

Band 27: F. Friedrich (2005)
Spectroscopic investigations of delaminated and intercalated phyllosilicates

Band 28: L. Niemann (2005)
Die Reaktionskinetik des Gipsabbindens: Makroskopische Reaktionsraten
und Mechanismen in molekularem Maßstab

Band 29: F. Wagner (2005)
Prozessverständnis einer Naturkatastrophe: eine geo- und hydrochemische
Untersuchung der regionalen Arsen-Anreicherung im Grundwasser West-
Bengalens (Indien)

Band 30: F. Pujol (2005)
Chemostratigraphy of some key European Frasnian-Famennian boundary
sections (Germany, Poland, France)

Band 31: Y. Dikikh (2006)
Adsorption und Mobilisierung wasserlöslicher Kfz-emittierter
Platingruppenelemente (Pt, Pd, Rh) an verschiedenen bodentypischen
Mineralen

Band 32: F. I. Müller (2007)
Influence of cellulose ethers on the kinetics of early Portland cement
hydration

Band 33: H. Haile Tolera (2007)
Suitability of local materials to purify Akaki Sub-Basin water

Band 34: M. Memon (2008)
Role of Fe-oxides for predicting phosphorus sorption in calcareous soils

Band 35: S. Bleeck-Schmidt (2008)
Geochemisch-mineralogische Hochwassersignale in Auensedimenten und
deren Relevanz für die Rekonstruktion von Hochwasserereignissen

Band 36: A. Steudel (2009)
Selection strategy and modification of layer silicates for technical
applications

Band 37: E. Eiche (2009)
Arsenic mobilization processes in the red river delta, Vietnam: Towards
a better understanding of the patchy distribution of dissolved arsenic in
alluvial deposits

Band 38: N. Schleicher (2012)
Chemical, Physical and Mineralogical Properties of Atmospheric
Particulate Matter in the Megacity Beijing

Band 39: H. Neidhardt (2012)
Arsenic in groundwater of West Bengal: Implications from a field study